From Noah to Italian Marxism to *Planet of the Apes*, Kowalewski assembles a dizzying array of philosophical, theological, and cultural texts in order to analyse our responses to eco-catastrophes. The resulting argument for an apocalypticism of the everyday contains no easy answers – only challenges to think and act differently.

Tommy Lynch, *Reader in Political Theology,*
University of Chichester, UK

Backed by a thorough study of the apocalyptic tradition, from Paul through Joachim da Fiore and Taubes to contemporary discourse of eco-apocalypse, Kowalewski proposes an aporetic apocalypticism as the only thought capable of taking us out of the current double bind in which the reproductive means of humankind's survival begin to endanger this very survival.

Agata Bielik-Robson, *Professor of Jewish Studies,*
University of Nottingham, UK

Resisting tired tropes and familiar narratives, Kowalewski brings together ancient and contemporary apocalypticisms into an exciting and provocative account of eco-politics in a time of crises.

Marika Rose, *Senior Lecturer in Philosophical Theology,*
University of Winchester, UK

A PHILOSOPHY OF CLIMATE APOCALYPTICISM

This book offers a long-overdue analysis of the ubiquity of eco-apocalypticism in current discourses on the climate crisis.

Drawing on a wide range of sources and theoretical traditions from ecological works and radical pamphlets, through political theology and continental philosophy to ancient and medieval apocalypses, the book sheds a comprehensive light on the concepts, processes, and experiences which circulate around the figure of the environmental end of the world. Importantly, this book argues that apocalypticism can provide a productive philosophical framework for addressing the climate catastrophe, enabling us to propose a distinctive answer to the fundamental question which haunts progressive ecological projects: how can we defend the world we find indefensible?

Appealing to students, academics, and researchers in philosophy, political theology, and environmental humanities, this book is a timely intervention which hopes to demonstrate that, when all else fails, it is the end of the world which may save the planet.

Jakub Kowalewski is Research Fellow at St Mary's University, Twickenham, London. He is also the editor of *The Environmental Apocalypse: Interdisciplinary Reflections on the Climate Crisis* (Routledge 2023).

Routledge Environmental Humanities

Series editors: Scott Slovic *(University of Idaho, USA)*,
Joni Adamson *(Arizona State University, USA)*
and Yuki Masami *(Aoyama Gakuin University, Japan)*

Editorial Board

Christina Alt, *St Andrews University, UK*
Alison Bashford, *University of New South Wales, Australia*
Peter Coates, *University of Bristol, UK*
Thom van Dooren, *University of Sydney, Australia*
Georgina Endfield, *University of Liverpool, UK*
Jodi Frawley, *University of Western Australia, Australia*
Andrea Gaynor, *The University of Western Australia, Australia*
Christina Gerhardt, *University of Hawai'i at Mānoa, USA*
Tom Lynch, *University of Nebraska, Lincoln, USA*
Iain McCalman, *Australian Catholic University, Australia*
Jennifer Newell, *Australian Museum, Sydney, Australia*
Simon Pooley, *Imperial College London, UK*
Sandra Swart, *Stellenbosch University, South Africa*
Ann Waltner, *University of Minnesota, US*
Jessica Weir, *University of Western Sydney, Australia*

International Advisory Board

William Beinart, *University of Oxford, UK*
Jane Carruthers, *University of South Africa, Pretoria, South Africa*
Dipesh Chakrabarty, *University of Chicago, USA*
Paul Holm, *Trinity College, Dublin, Republic of Ireland*
Shen Hou, *Renmin University of China, Beijing, China*
Rob Nixon, *Princeton University, Princeton NJ, USA*
Pauline Phemister, *Institute of Advanced Studies in the Humanities, University of Edinburgh, UK*
Sverker Sorlin, *KTH Environmental Humanities Laboratory, Royal Institute of Technology, Stockholm, Sweden*
Helmuth Trischler, *Deutsches Museum, Munich and Co-Director, Rachel Carson Centre, Ludwig-Maximilians-Universität, Germany*
Mary Evelyn Tucker, *Yale University, USA*
Kirsten Wehner, *University of London, UK*

The *Routledge Environmental Humanities* series is an original and inspiring venture recognising that today's world agricultural and water crises, ocean pollution and resource depletion, global warming from greenhouse gases, urban sprawl, overpopulation, food insecurity and environmental justice are all *crises of culture*.

The reality of understanding and finding adaptive solutions to our present and future environmental challenges has shifted the epicenter of environmental studies away from an exclusively scientific and technological framework to one that depends on the human-focused disciplines and ideas of the humanities and allied social sciences.

We thus welcome book proposals from all humanities and social sciences disciplines for an inclusive and interdisciplinary series. We favour manuscripts aimed at an international readership and written in a lively and accessible style. The readership comprises scholars and students from the humanities and social sciences and thoughtful readers concerned about the human dimensions of environmental change.

A Philosophy of Climate Apocalypticism
In and Against the World
Jakub Kowalewski

For more information about this series, please visit: www.routledge.com/Routledge-Environmental-Humanities/book-series/REH

A PHILOSOPHY OF CLIMATE APOCALYPTICISM

In and Against the World

Jakub Kowalewski

Routledge
Taylor & Francis Group
LONDON AND NEW YORK

from Routledge

Designed cover image: © Piotr Miemiec
Description: The "Krystyna" tower, Bytom, Poland.

First published 2025
by Routledge
4 Park Square, Milton Park, Abingdon, Oxon OX14 4RN

and by Routledge
605 Third Avenue, New York, NY 10158

Routledge is an imprint of the Taylor & Francis Group, an informa business

British Library Cataloguing-in-Publication Data
A catalogue record for this book is available from the British Library

ISBN: 978-1-032-39129-8 (hbk)
ISBN: 978-1-032-39126-7 (pbk)
ISBN: 978-1-003-34851-1 (ebk)

DOI: 10.4324/9781003348511

Typeset in Sabon
by Apex CoVantage, LLC

CONTENTS

ACKNOWLEDGEMENTS

In a recent documentary about his life, Arnold Schwarzenegger makes it clear that he was not a self-made man and that his achievements had been possible only because of other people. Although writing a book feels like a Herculean effort, it certainly can't be compared to winning Mr. Olympia bodybuilding competition; nonetheless, this work is a culmination of decades of support, encouragement, and mentorship I received from others.

I would like to thank my schoolteachers – Carmody Grey, Alistair McConville, and Rupert Crisswell – who, perhaps unwittingly, introduced me to philosophy as a coincidence of opposites: simultaneously the most serious, scholarly, and existential pursuit with irreversible, life-altering consequences and the most playful, irreverent, fun, and (nearly) meaningless pastime.

I was lucky enough to attend the University of Sussex, where I was taught by brilliant philosophers. I would like to thank Tanja Staehler in particular. Our seemingly inconsequential conversation about a summer research project on Derrida and Husserl turned into years-long intellectual adventure and an inescapable point of reference for everything I've written since.

During my PhD at the University of Essex, I interacted with some truly remarkable thinkers, including my fellow doctoral students – thank you for the countless hours you spent discussing philosophy (and related life problems) with me. Throughout my time at Essex, I worked with the most incredible supervisor: Irene McMullin. I am so grateful for the opportunity to think, talk, and write with such an impressive and inspiring *Doktormutter*. Thank you Irene.

The nightmarish experience of the postdoctoral period was in part counteracted by working alongside amazing colleagues at the University of Winchester, University of Chichester, and Kaplan International College London.

In particular, I would like to thank Timothy Secret, whose professional support and intellectual influence helped me not only to survive on the academic job market but also to find solutions to theoretical problems which haunted this project.

I would like to thank my colleagues who, over the last couple of years, created a supportive, enabling, and validating work environment conducive to writing (and finishing!) a book: Andrew Jones and Rose Taylor from Linking London, as well as Roland Daw and the colleagues from the Guardians of Creation Project: Emma Gardner, Edward de Quay, and John-Paul de Quay.

Huge thanks to the editors at Routledge – Grace Harrison, Matthew Shobbrook, and Radhika Gupta. Matt in particular displayed a saintly patience when responding to my emails and never-ending requests for deadline extensions.

Massive thank you to my family – especially my parents Beata and Jacek – for their ongoing support and for all the food and drink that help me to regularly forget about the end of the world. Thank you to all my friends; our time together always beautifully counterbalances the weight of everyday existence (even if I put a few of you to sleep when talking philosophy).

Finally, I would like to thank Kimia Gashtili – who, at the time of writing these words, is pregnant with our first child – *for everything*.

INTRODUCTION AND INITIAL HYPOTHESES

It has become customary to begin a book on environmental crisis with a long list of climate change-related disasters (floods, droughts, wildfires, the raising of sea levels, the disappearance of species and ecosystems, etc.). Such a practice has obvious advantages – it situates a given work in the midst of a contemporary problem, it endows a book's argument with a sense of urgency, and it generates the rhetorical force behind author's prose.

I must confess that I was also tempted by this (virtually mandatory) opening. However, compiling an inventory of climate catastrophes proved impossible – there were simply *too many* instances. Any selection seemed arbitrary and, frankly, unjust, rendering visible some disastrous events only on the conditions of making countless others invisible.

Importantly, beginning this work with an enumeration of environmental catastrophes seemed inappropriate for another reason. This book is strictly speaking not about climate change-related disasters. It is, rather, about the apparently ubiquitous designation of the environmental crises as a *climate apocalypse* (at least in Western popular and scientific discourses). Of course, on the one hand, *climate apocalypse* is simply a synonym for today's catastrophic situation structured by disastrous events and processes; on the other hand, however, the term *apocalypse* introduces something extra to the situation it names, a certain semantic excess which cannot be explained away simply as a rhetorical device.

The two intuitions which motivate this book are, first, that additional meanings are smuggled in whenever the term *climate apocalypse* is used; and, second, that these implied meanings point to a whole set of unarticulated philosophical commitments. The present work, by attempting to articulate the theoretical and ethico-political effects of apocalypticism, situates itself in

DOI: 10.4324/9781003348511-1

the discursive gap which opens up between climate change-related disasters and their designation as *climate apocalypse*.

What can be found if this excessive gap opened by *climate apocalypse* is prodded and investigated? This introduction proposes a number of hypotheses which are taken up, either directly or indirectly, in the rest of the book.

A contested notion

Sometimes, to understand a contested term we need to understand the reaction it provokes. In the case of disputed or contentious notions, the very utterance of the word or a phrase is inseparable from a value judgement, often present as an affective inflection with a power to determine the meaning of the word. Paradoxically, our response to a term comes *before* its expression, and it is this prior reaction which moulds the concept we state.

The phrase *climate apocalypse* (and its cognates: *environmental apocalypse* and *eco-apocalypse*) undoubtedly belongs to the category of contested notions. Its mention in everyday conversations, newspaper articles, political pamphlets, or academic publications is almost always shaped by a prior reaction of the author.

For many, climate apocalypse is too alarmist, too pessimist, too quietist, too apolitical, too Western-centric, or too entwined with religion to be useful for a serious climate discourse (and this is only a tip of an iceberg of criticisms waged against this term).

For others, referring to a set of climate-related disasters as *eco-apocalypse* offers an evocative way to frame one's critical analysis of climate change. Consequently, placing the environmental crisis in an apocalyptic perspective can become a chosen rhetorical tactic. However, in those instances the term *apocalypse* functions either as a vivid metaphor or in its original Greek meaning, setting up a *revelation* promised by the author.

For a few, we are living through a *literal* climate apocalypse – if you have ever tried to take a walk during a heatwave, you know what the end of the world feels like. In fact, for those few authors, the equation between the climate catastrophe and apocalypse seems so obvious that its evocation appears trite. Climate change is simply another reiteration of a long chain of actual or potential "ends of the world," which routinely make up human history (e.g. political upheavals, financial crises, pandemic, wars, natural disasters). Here, the literal use of *climate apocalypse* is allowed only on the condition of asserting its ordinariness and repetitiveness, thus deflating the affective force of the hyperbolic notion of "the end of the world."

This book's understanding of *climate apocalypse* finds itself the closest to the last position. But, in contrast to the rather jaded apocalypticism outlined earlier, I believe that both the *obviousness* of the environmental apocalypse and the *repetitive character* of "the ends of the world," far from deflating apocalyptic notions, offer clues to unpack the rich philosophical presupposition of the eco-apocalyptic discourse. These presuppositions, in turn, can explain both why *climate apocalypse* can generate a whole host of (often incompatible) criticisms and why it can be appropriately used in its metaphorical and etymological senses. In other words, it is when the contested notion seems the most familiar that its strange theoretical elements come to the surface, ready to be articulated. Let's begin by looking closer at both the obviousness of climate apocalypse.

Obvious, too obvious

In 1936, the phenomenologist Eugen Fink noted our seemingly intimate acquaintance with both our conscious experience and the meaning of the term *consciousness* – "this familiarity," Fink tells us, "gives rise to the illusion that consciousness is something immediately given." For Fink, however, phenomenology, by uncovering the "overlooked" structures that underlie our conscious experience of ourselves and the world, demonstrates that our immediate acquaintance with consciousness hides more fundamental meanings. Consequently, the task of phenomenology is "to loosen the rigid fundamental posture of our naïve-natural life-world attitude (living straightforwardly towards things)" and to critique "the *illusion* of everyday, given immediacy" (Fink 1970, 385–387).

Interestingly, Fink's phenomenological critique of "naïve and dogmatic" (Fink 1970, 387) attitudes stemming from the obviousness which characterises our everyday experience is echoed by the French Marxist Louis Althusser in his theory of ideology. Thirty years after Fink, Althusser writes:

> Ideology is so much present in all the acts and deeds of individuals that it is *indistinguishable from their "lived experience,"* and every unmediated analysis of the "lived" is profoundly marked by the themes of ideological obviousness. When he thinks he is dealing with pure, naked perception of reality itself, or a pure practice, the individual (and the empiricist philosopher) is, in truth, dealing with an impure perception and practice, marked by the invisible structures of ideology; since he does not *perceive* ideology, he takes his perception of things and of the world as the perception of "things themselves," without realizing that this perception is given him only in the veil of unsuspected forms of ideology.
>
> *(Althusser 2011, 25–26)*

Fink's and Althusser's respective suspicions regarding obviousness of experiences, concepts, and practices enable us to propose the following hypotheses: first, appearing with a straightforward familiarity is an effective way for an experience (or concept, or practice) to hide its own conditions; second, since as Althusser tells us, everyday obviousness is generated by the work of ideology, that which an obvious experience (or concept, or practice) makes invisible is its own ideological character. In other words, when a phenomenon, like a policeman, shouts: "You know exactly what's going on here, move along!" – it is at this point that the phenomenon works the hardest to cover over the traces of the ideological structures which make it possible. Consequently, the phenomenological task, recommended by Fink, of disturbing the naivety of living straightforwardly towards things and notions by uncovering the overlooked conditions of phenomena is inseparable from an analysis of ideology and its effects.

If we apply these conclusions to the obviousness with which the environmental catastrophe appears as an apocalypse, we can suggest that this obviousness indicates the work of ideological dissimulation, rendering invisible the constitutive structures of the environmental apocalypse. In short, the more everyday the notion and experience of climate apocalypse seem, the closer we are to its ideological character.

Ideology and eco-apocalypse

For Althusser, the role of ideology – in tandem with economy and politics – is to make possible the reproduction of our conditions of existence (Althusser 2014). Now, if discourses operating with the notion of eco-apocalypse possess ideological qualities, then they would also play a part in ensuring the possibility of reproduction.

Admittedly, it does sound rather odd to propose that the function of climate apocalypse – a term designating none other than the "end of the world"! – is to secure any form of *reproduction*, especially the reproduction of society's conditions of existence. Yet a closer look reveals that the desire to protect our earth, and consequently earth-bound species, is central to the logic of most (though, as I show later, not all) discourses evoking the apocalyptic framing of the environmental crisis.

Climate apocalypse is often evoked *critically* as a tool highlighting how current forms of reproduction (ideological, political, economic etc.) lead towards a literal "end of the world" for both humans and other species. As Timothy Secret puts it:

> When we think through climate change, we realise that those acts of economic and social reproduction that we had been engaging in to preserve the world against the external forces of death and decomposition are themselves directly leading us towards the end of the world.
>
> *(Secret 2023, 216)*

Eco-apocalypse therefore names a paradoxical situation where the very ideas, practices, and processes that ensure society's reproduction threaten society's environmental conditions of existence.

Importantly, the critical type of eco-apocalyptic discourses often advocate for an *end* to unsustainable reproductive practices and the representations which sustain them. Critical apocalypticism does so to protect the environmental conditions of reproduction, which ensure that planet earth remains habitable for humans and other species. The structure of this argument can be illustrated with a practical example – choosing not to have children for environmental reasons. Elizabeth Pyne explains:

> In the contexts where lifestyles are carbon-intensive, recent climate commentary includes think pieces and op-eds – often anguished, sometimes angry, occasionally smug – that adapt the sentiment. . . . "I'm choosing not to have kids because I care about the environment." This genre extends in fascinating ways into the youth movement. In 2019, 18-year-old Canada-based activist Emma Lim pledged not to have kids until governments begin taking meaningful action on climate change; several thousand people signed on, using the hashtag #NoFuture-NoChildren.
>
> *(Pyne 2023, 35)*

The conscious decision to stop procreating – that is to say, to stop humanity's reproduction – has an apocalyptic air about it. When universalised, it would amount to the eventual extinction of the human species. Of course, as Pyne makes clear in the passage earlier, the environmental choice not to have children is usually framed as a less radical, first-personal declaration whose goal is either to reduce the carbon footprint of one's family unit or to influence policymakers by exploiting their attachment to human reproduction. Here, the value of reproduction which underlies the potentially apocalyptic decision becomes palpable. What motivates some people to stop procreating is the recognition that *human* reproduction on the current scale is both incompatible with and secondary to *environmental* conditions of reproduction. Therefore, the goal of not having children, far from any apocalyptic connotations with non-reproduction, is to mitigate the effect of climate change (either individually or politically) and, in so doing, ensure the optimal environmental possibility of human and non-human *reproduction*. Even in a more extreme hypothetical case, where *most* of us decide to reduce the size of the human species, this Christ-like self-sacrifice would be desirable only on the condition that it "saves the planet," that is it secures the continual reproduction of life on earth, including the remaining human society.

Thus, we can suggest the function of the figure of apocalypse in critical apocalypticism is to correct how society reproduces itself to prevent the world from ending and to ensure the self-preservation of society.

There is something common-sensical about using the figure of "the end of the world" to reduce or even end *some* reproductive practices to protect our environmental conditions of existence more effectively. However, one of the consequences of the critical eco-apocalyptic logic is that the defining quality of climate apocalypse – its "end-of-the-world-ness" (if I may be forgiven this clunky expression) – is neutralised and placed in the service of the continual existence of the world. Speaking figuratively, "the end of the world" in critical apocalypticism becomes the paradoxical way in which the world expresses its need for survival, encouraging us to avert the destruction of the planet.

This, in turn, means that in the critical eco-apocalyptic logic examined earlier, climate apocalypse is subordinated to its ideological function: safeguarding the environmental possibility of the reproduction of life. By identifying self-destructive practices which threaten the environmental conditions of human and non-human existence, critical climate apocalypticism helps dominant ideologies, and the reproductive processes they mediate, to come face to face with their weak points, to fix them, and consequently, to succeed in their goal of ensuring the reproduction of society in an optimal environment. In other words, although critical eco-apocalypticism often claims to *oppose* dominant ideologies and practices, its actual role is to *improve* them. The interventions called for by critical eco-apocalypticism – to correct reformable mistakes and to stop unreformable errors – allow dominant ideologies and the concomitant practices to perform their reproductive function more successfully.

Certainly, one may argue that placing the often-unhealthy obsession with annihilation characteristic of apocalypticism in the service of planet-affirming goals turns climate apocalypse into a theoretically responsible notion with a potential to adequately address the problems of climate change. The destructive force of "the end of the world" becomes domesticated, making climate apocalypticism socially acceptable because it is useful from the point of view of society's reproduction. But to present eco-apocalypticism as a critical yet ultimately safe ideological standpoint is to flatten out the multidimensional phenomenon of "the end of the world" and to suppress the radical potential of apocalyptic consciousness. As I will show in the next section, it is possible to *re-wild* climate apocalypse, revealing the multifaceted "end-of-the-world-ness" overflowing its narrow ideological confines.

The strategy of refusal

The work of Thomas Lynch can be read as an attempt to think apocalyptically *against reproduction*. Lynch's project relies on two interconnected forms of refusal: first, the refusal of the world in its entirety; and second, the refusal of affirmative proposals offered as an alternative to the world.[1]

In his piece aptly titled "How I Learned to Stop Hoping and Hate the World," Lynch writes:

> The world was made through violence, disease, destruction, extraction and pollution. The evils of colonialism, slavery and climate change are not incidental to the world. They have been and are what makes the world. . . . The world cannot be put right or fixed, nor can it be undone so that we might return to some imagined ideal era. This point is at the essence of apocalypticism.
>
> *(Lynch 2022)*

The indefensible ethico-political character of the world, Lynch tells us, necessitates the *negation of the world in its totality*. It is not only the case that the world is not worth fighting for; rather, the world should be actively fought against. Importantly, for Lynch, the struggle against the world and the desire for its end shouldn't be fuelled by any positive images of a better future. Insofar as visions of an alternative future are produced within the world, they reproduce the flaws which define their worldly point of origin. Lynch's apocalypticism, therefore, "refuses the blackmail of affirmation from the outset . . . it only offers a refusal" (2019, 130).

Now, the apocalyptic negation proposed by Lynch can be interpreted in two ways: less and more radical. The negation of the world could be understood either as a negation of the totality of reproductive practices or as a negation of the totality of reproductive practices *and* their environmental conditions.

On both readings, Lynch's radical apocalyptic position can be contrasted with critical apocalypticism – the latter's goal is to end only *some* forms of reproduction rather than a totality of social practices. However, it is only on the second, more extreme reading, which negates also the environmental conditions of reproductive practices, that Lynch can truly do away with an attachment to reproduction.

A critic may argue, first, that Lynch's refusal of any affirmative proposals is unhelpful from a political point of view. Even if we agree with the less radical formulation of Lynch's apocalypticism, how can we struggle against societal practices without some images of the future we want to bring about? It is counter-intuitive, if not outright impossible, to design political strategies without an idea of the desired outcomes. And since Lynch's apocalypticism cannot be translated into the language of demands and solutions, his project would appear apolitical and thus useless for climate politics. Second, in its more radical formulation, the negation of the world is a nihilistic dead end, especially when confronted with the climate crisis. If apocalypticism gives up on our environmental conditions of existence, it is a standpoint that goes against the core commitment of any ecological project.

This is too high of a price to pay, even when confronted with the world's cruelty and injustice.

To address this criticism, it is helpful to read Lynch alongside the Italian Marxist Mario Tronti. In his *Workers and Capital*, written 60 years ago, Tronti puts forward arguments which anticipate Lynch's apocalypticism. Tronti's anti-capitalism also refuses any positive visions of a better world.

> Indeed, none of what stands in front of us today is the future. And to base the model for a future society on the analysis of present-day society is a bourgeois ideological vice. . . . No worker who is fighting against a boss is going to ask, "And then what?" The fight against the boss is everything. . . . No, the problem today is not a matter of envisioning what ought to replace the old world; we are still facing the issue of how to tear it down.
>
> *(Tronti 2019, xxiv–xxv)*

Tronti's negativist political strategy centres on the refusal of work. To stop working is to halt the reproduction of capitalist economy, but it is also to undermine the very conditions which make possible the existence of the working class. Tronti makes clear that to "create a point of rupture" with the present, the working class must pass through the paradoxical standpoint of attacking its own conditions of reproduction. "To fight against capital, the working class must fight against itself *qua* capital. . . . A working-class struggle against work, the worker's struggle against her own condition as a wage-labourer" (2019, 273).

The formal similarity between Tronti's and Lynch's respective arguments can help us to make explicit the political potential of the latter: to refuse reproduction is the necessary condition of transformative political action; furthermore, this refusal does not have to rely on images of the future to be effective. According to both Tronti and Lynch, destroying the present is *enough* from the point of view of both theory and political strategy (Lynch 2019, 130). In the words of Tronti, "Working-class science is but a means for the organisation of this destruction . . . and that's just fine" (2019, 271).

While I am sympathetic to Lynch's (and Tronti's) project of espousing *non-reproduction*, I believe that it must be qualified in two ways. First, radical apocalypticism which advocates for non-reproduction is *contradictory* because the end of the world always involves a degree of world-creation, which enables the reproduction of our conditions of existence (albeit in a transformed way). This intuition can be found expressed at different points by both Tronti and Lynch. For Tronti, "a block on the present" is "an instance of its possible reorganisation" (2019, 261); for Lynch, the end of the world "is the possibility of other possibilities" (2019, 32). In consequence, the difference between critical and radical apocalypticism becomes one of *emphasis*:

while the former subordinates ends to a more fundamental demand for reproduction, the latter in negating the world inadvertently smuggles in elements capable of reproducing our conditions of existence. Importantly, as I will discuss in more detail later, it is precisely this internal flaw of radical apocalypticism – the fact that its programmatic concern with non-reproduction hides and perpetuates its supposed enemy, that is the ideological need for reproduction – which allows it to escape the charge of nihilism and which makes calls of the refusal of the world compatible with ecological aims.

Second, I do not think that it is feasible to refuse the future on apocalyptic grounds. Apocalypticism is a historical phenomenon; the study of past ends of the world can help us to identify apocalyptic tendencies which organise the present and which *point to the future*. In other words, while Lynch implies that the end of the world is a rupture which separates history into two – the negated present and the unknown future – I would like to suggest that the end of the world (in addition to functioning as a break) also establishes a historical connection between the past, present, and the future. Note however that future here figures not as a utopian image of new society (which, as Tronti tell us, can only be ideological) but as a consequence of processes set in motion by past apocalypses.

In the reminder of this introduction, I will elaborate on the two qualifications to radical apocalypticism sketched above.

The triple power of the apocalypse

Intuitively, the apocalypse presupposes the existence of the world as its condition: for "the end of the world" to take place, *there must be a world* which precedes and makes possible the apocalypse. This intuition is operative in how we conceptualise the environmental "end of the world" in the West: we tend to think of climate apocalypse as either a future outcome of or a present malfunctioning of society's reproduction, that is as a secondary modification or a pathological flaw of the *more primary* forms of world's reproduction.

I would like to propose that we turn this intuition on its head: my hypothesis is that *the end of the world can be prior to the existence of the world*. Such a claim implies another one: that of the existence of a *plurality* of worlds.

The belief in multiple worlds and apocalypses can be found expressed, for example, in the traditional division between the "Old World" and the "New World." Importantly, for our purposes, the history of Europe and its colonies demonstrates, alongside the world-destroying power of the apocalypse, the *world-forming* function of apocalyptic ends.

The name New World insidiously covers over the apocalyptic reality of the "discovery" of the Americas. By thinking of the symbolic date of 1492 as a moment at which a new world is *born* on another continent, modern European consciousness was able to simultaneously block out and redeem the *destruction* of the Amerindian universe which followed. The progressive narrative about the "discovery" of the New World by the Old World obscures the fact that, as Déborah Danowski and Eduardo Viveiros de Castro note, the arrival of European colonisers marks "the end of the world" for the native inhabitants of the Americas:

> The indigenous population of the continent, larger than that of Europe at the time, may have lost – by means of the combined action of viruses (smallpox in particular being spectacularly lethal), iron, gunpowder, and paper (treaties, papal bulls, royal *encomienda* concessions, and of course the Bible) – something of the order of 95 percent of its bulk throughout the first one and a half centuries of the Conquest. That would correspond, according to some demographers, to a fifth of the planet's population. We could therefore call this American event the First Great Modern Extinction, when the New World was hit by the Old one as if by a giant celestial body.
>
> *(Danowski & Viveiros de Castro 2016, 104–105)*

The repression of the colonial apocalypse is further solidified by the dissimulating work of the term the Old World. Contrary to what the term indicates, Europe does not possess an ontological priority in relation to its American "discovery." In fact, it is the New World – or more accurately, its destruction – which makes possible the emergence of the European continent in a new form:

> The genocide of Amerindian peoples – the end of the world for them – was the beginning of the modern world for Europe: without the despoiling of the Americas, Europe would have never become more than the backyard of Eurasia. . . . No pillage of the Americas, no capitalism, no Industrial Revolution, thus perhaps no Anthropocene either.
>
> *(Danowski & Viveiros de Castro 2016, 107)*

The analysis of the pair of notions New World–Old World reveals a twofold distortion: on the one hand, the narrative about the birth of the New World downplays – to a point of obscuring – the end of the world inseparable from Columbus's "discovery." On the other hand, the supposed anteriority of the Old World masks the generative role of the destruction of the indigenous world for Europe – the end of the world for the native people of the Americas is the beginning of the modern European universe.

Furthermore, even though both the Americas and Europe are worlds constituted by the power of the apocalypse, the world-annihilation/world-creation processes are disturbed unevenly, accounting both for the specificity of the post-apocalyptic continents and for the unequal relation between these end-worlds. Consequently, we can suggest that in addition to its capacity to destroy and engender worlds, the colonial apocalypse possesses a power to *maintain* a specific arrangement of end-worlds.

The triple power found in the historical example of the colonial apocalypse can be generalised and ascribed to apocalypse *as such*. The end of the world is not only the possible effect of worldly processes; it is also a *cause* responsible for the emergent conditions of existence. In the words of Taubes: "The apocalyptic principle combines within it a form-destroying and a forming power. Depending on the situation and the task, only one of the two components emerges, but neither can be absent" (Taubes 2009, 10). Additionally, apocalypses stabilise the relationships between ruined and emergent worlds, and as such they play the role not only in the destruction and creation but also in the maintenance and the reproduction of our conditions of existence.

However, we should be careful not to interpret the apocalyptic world-forming and world-maintaining as in any way justifying or redeeming the ends of the world. Although culturally we may be used to believing in "the miracle of birth," the apocalyptic emergence of another world may not be distinguishable from the post-apocalyptic ruins, and any redemptive possibilities which this "birth" may (or may not) harbour are inseparable from the irredeemable trauma of destruction.

At this point, we can propose the following definition of the apocalypse: the end of the world stands for any practices, events, or processes which threaten, limit, or render impossible the reproduction of individual, collective, or planetary life and which, in so doing, enable the emergence and the stabilisation of new conditions of existence. Note that this definition implies that an *absolute end* is impossible (at least on a planetary scale) – the triple power of the apocalypse necessitates world-creation and world-maintenance, even if only minimally.

This definition would suggest that both history and the present are populated with multiple examples of possible and actual ends of the world. This insight is often obscured by a belief that there is only one world, which can be annihilated by a single apocalyptic event taking place in the future. Apocalypticism, therefore, can offer a useful theoretical lens through which to study the past, the present, and the future: whenever we witness an advent of a new world, we can anticipate it to be an effect of the constituting work of the apocalypse; conversely, whenever a world ends, we can expect it to generate a new world.

Furthermore, the triple power of the apocalypse can help us to explain why Lynch's radical apocalypticism – which negates both the world and its conditions of existence – cannot be classified as nihilistic. If, as I mentioned earlier, an *absolute* end is impossible because every world-destruction implies world-constitution and world-maintenance, then even the total negation of the world presupposes an emergence of new conditions of existence and forms of reproduction.

Lynch, however, is reluctant to offer any positive vision of the consequences of the end of the world. Here, I depart from Lynch. I believe that an examination of past and present effects of apocalypses can give us access to apocalyptic tendencies – historical developments which extend to the future by pointing towards specific forms of the destruction, survival, and generation of societies' conditions of reproduction. This knowledge, in turn, can inform political action, including climate politics.

To examine the historical character of apocalypticism, in the next section, I will consider in more detail the quality of apocalypticism which can be found alongside its obviousness: the repetition of apocalypticism across history.

The end is near, yet again

A cursory study of the history of apocalypticism reveals the persistence of apocalyptic consciousness across millennia. In fact, even one generation can experience multiple premonitions of the end. In *Spectres of Marx*, published in the wake of the collapse of the Soviet Union – a historical period dominated in the West by the apocalyptic theme of the "end of history" – the French philosopher Jacques Derrida reflects on the parallels between the discourses of the 1990s and 1950s. Echoing our earlier observations, Derrida presents apocalypticism as both immediately familiar and trite:

> [T]he eschatological themes of the "end of history," of the "end of Marxism," of the "end of philosophy," of the "ends of man," of the "last man" and so forth were, in the '50s, that is, forty years ago, our daily bread. We had this bread of apocalypse in our mouth naturally, already, just as naturally as that which I nicknamed after the fact, in 1980, the "apocalyptic tone of in philosophy". . . . Thus for those with whom I shared this singular period . . . the media parade of current discourse on the end of history and the last man look most often like a tiresome anachronism.
> *(Derrida 1994, 16)*

In keeping with Derrida's ghostly vocabulary, we can suggest that apocalypse is a spectre that haunts Western history with a rather boring regularity. However, we can ask – what specifically is repeated across 1950s and 1990s apocalypticism? As Derrida notes, the apocalyptic ghost visiting 1950s France possessed a body constituted by the "historical entanglement" between philosophy and political events, reminiscent of its 1990s counterpart:

> [O]n the one hand, the reading or analysis of those whom we could nickname the *classics of the end*. They formed the canon of the modern

apocalypse (end of History, end of Man, end of Philosophy, Hegel, Marx, Nietzsche, Heidegger, with their Kojevian codicil and the codicils of Kojève himself) . . . *on the other hand and indissociably*, what we had known or what some of us for quite some time no longer hid from concerning totalitarian terror in all the Eastern countries, all the socio-economic disasters of Soviet bureaucracy, the Stalinism of the past and the neo-Stalinism in process (roughly speaking, from the Moscow trails to the repression in Hungary, to take only these minimal indices).

(1994, 16)

The apocalypticism of the 1990s seems like a *déjà vu* for Derrida because the failure of Soviet communism and the concomitant pertinence of philosophies of the end of history were the daily bread of Derrida's generation already in the 1950s. In other words, it is the historical entanglement of philosophy and political events which is responsible for the repetition of 1950s apocalypticism in the 1990s and for its experience as a "tiresome anachronism."

It strikes me that in his unenthusiastic reflections on the analogies between the 1950s and 1990s, Derrida stumbled across something that may be called a *form* of apocalypticism: it is possible to compare and even superimpose the apocalypticisms of the 1950s and 1990s, because their respective contents are structured by the shared form constituted by the historical entanglement of political events and philosophy; it is the apocalyptic form which is responsible for the sense of familiarity and repetitiveness reported by Derrida.

What I would like to propose is that if climate apocalypticism is another reiteration of the tired theme of "the end of the world," this is because it also shares a form with other types of apocalyptic discourses which preceded it or surround it: the historical entanglement between a political event – in this case, the environmental crisis – and theoretical discourses on the end (e.g. of certain reproductive practices).

The last claim introduces an important point into our discussion. If we want to maintain that each reappearance of apocalypticism amounts to a *déjà vu*, it is only on the condition that we emphasise the shared form but ignore the – however slight – differences of content. Kojève is not Fukuyama, even if both discuss "the end of history"; the failure of Stalinism is not the same as the collapse of the Soviet Union, and so on. The specificity of apocalyptic contents is even more visible when we compare the 1990s discourse on the end of history with today's environmental apocalypticism. In short, apocalypses are repetitive because of their form but remain individuated because of their contents. The latter function as a tissue – in each case specific to a given historical context – filling out and covering the generic skeleton of form. Similarly to the creature from Andrzej Żuławski's film *Possession* (1981), the apocalyptic spectre needs to parasitically feed on its surroundings to take shape.

Prophets or subject positions

The image of apocalypses as creatures which swallow historical material they encounter captures figuratively the apocalyptic atmosphere specific to certain contexts, for example Derrida's milieu in the 1950s or modern-day Western environmentalism. In such situations, the appropriateness of apocalyptic language becomes self-evident, as if suggested by "things themselves."

However, to emerge, any type of apocalypticism requires *formulation*: a selection and linking together of relevant political events and theoretical sources. This selection is made by a *subject* who systematises (more or less effectively) the underdetermined apocalyptic atmosphere in an overdetermined historical situation.

This role has historically been played by a prophet. Although individual prophets claim to be concerned with the seemingly universal problem of "the end of the world," they nonetheless select, even if not consciously, which theories and political events are most pertinent for their articulation of the apocalypse. This is particularly noticeable in the case of environmental apocalypticisms, where the planet-wide problem of the ecological apocalypse is nonetheless framed in terms of *localisable* theoretical and historical material, often reflective of the biases of the author. As Spinoza's observes, and his insights apply also to today's prophets of the apocalypse, revelation "varied from one prophet to another . . . it depended upon the disposition of his bodily temperament, his imagination, and the belief he had previously adopted" (Spinoza 2007, 30).[2] Apocalypticism, therefore, presupposes what we can call *subject positions* responsible for the (conscious and unconscious) selection of the material – localisable features reflective of the prejudices, characteristics, and the milieu of any "prophet." Subject positions would therefore be synonymous with interior and exterior *world* of the prophet, a familiar context which determines the content of a given apocalyptic prophecy.

Far from being a flaw, the possibility of making the apocalypse *one's own* explains the vivacious persistence of apocalyptic consciousness, which resurfaces renewed in different historical contexts. As Jacob Taubes rightly observes, "if they are to retain their validity in the community, apocalypses have to be adapted to each new situation, and this inevitably leads to interpolations" (Taubes 2009, 70).

It is worth noting that some subjects are drawn into the orbit of apocalypticism *without understanding* or *unconsciously*. Some thinkers formulate apocalyptic arguments while not realising they are doing so; others offer insights which, unbeknownst to them and independently of their intentions, can be made to speak to an apocalyptic problematic. In fact, we can find an example of such an unwitting apocalypticism already in the Bible: the prophet Daniel "who not only failed to understand what God said to him (the lot of all prophets) but even the explanation provided for him!" (Althusser 1993, 216).

Apocalyptic hermeneutics: concordances and historical blocks

At this point, I would like to say a few things about the method assumed in this work. I take seriously Emmanuel Levinas's belief that methodological transparency is hard to achieve and that those "who have worked on methodology all their lives have written many books that replace the more interesting books that they could have written" (Levinas 1998, 89). With this in mind, I shall keep my remarks brief.

We can study "the end of the world" by mapping the similarities and differences between individual expressions of apocalypticism across history. I refer to this tracking of apocalyptic relations in history as a *study of concordances*.

The study of concordances presupposes two types of historical connections: the *more visible* relations, presupposing chronological and linear time, and the *less visible* – but no less interesting – relations which cuts across linear chronology. For example, we can represent the relationship between the respective apocalypticisms of ancient apostle Paul, the medieval monk Joachim of Fiore, and the modern philosopher Hegel as separated by thousands of years of political events and epochal shifts in Western thought. However, as Taubes shows, a closer study of the three thinkers reveals intimate parallels – or concordances – between their respective ideas (2009). This would suggest that the formulations of Paul or Joachim are in an important sense closer to Hegel's than to some of the latter's own contemporaries, especially those ignorant of the apocalyptic atmosphere which surrounds them. By studying concordances, we can construct connections and distances between historical expressions of apocalypticism not reducible to chronological history.

An alternative apocalyptic hermeneutic would, by contrast, emphasise *historical blocs* – shared modes of thought responsive to political events within the same moment in time. Derrida intuits such a bloc when he describes the apocalypticism of the 1950s, a general atmosphere where politics and philosophy reflected share apocalyptic anxieties. Historical blocs are normally designated with umbrella terms, such as *nuclear* apocalypse or *financial* apocalypse, gathering together various apocalyptic formulations found within a specific historical moment. My contention is that *climate apocalypse* is one of such terms capable of designating a particular historical bloc, that is to say, a specific apocalyptic atmosphere defining the current historical context.

In practice, however, the two ways of reading the "end of the world" – focusing on concordances or historical blocs respectively – are inseparable. The study of concordances finds hidden connections cutting across chronological history populated with historical blocs. Coming into contact with concordances, in turn, deconstructs historical blocs – the latter can shrink or expand thanks to often-surprising political and theoretical connections introduced by the study of concordances. To give an example: the concordances between Paul, Joachim, and Hegel are interesting not only because they seem relatively autonomous in relation to chronological history; their power lies in the fact that they can be made to bear on a formulation of a particular historical bloc – such as the one designated by the umbrella term *climate apocalypse* – setting the reformulation of the present moment in motion.

In and against the world

Is it possible to combine apocalypticism which fights against our conditions of existence with environmentalism whose goal is to ensure the protection of those very conditions? How can we reconcile the fact that the world is morally and politically indefensible with calls to protect our planet? Can any pro-environmental tactics be derived from the apocalyptic strategy of refusal?

If Lynch's radical apocalypticism could be informatively paired with Tronti's *Workers and Capital*, the basic tenets of the apocalypticism I develop in this book can be articulated more fully when read alongside another political text: *In and Against the State* written by the London Edinburgh Weekend Return Group (LEWRG) in the late 1970s. It strikes me that *In and Against the State* expresses and addresses – albeit in a different language and within a different problematic – the paradoxes generated by an attempt to marry apocalypticism with environmentalism.

In and Against the State asks the following question: how should socialist public sector workers respond to cuts to state funding, including the loss of state jobs? On the one hand, such cuts, and the redundancies they involve, threaten the means of existence of people who rely on their employment in state institutions. This reality generates "the need to defend the state." On the other hand, the same public sector workers "do not feel that it is 'their' state and know that the state itself oppresses them" (1980, 6). The dilemma experienced by a socialist public sector worker mirrors the situation of anyone who finds the world alienating and oppressive but who nonetheless experiences the need to protect our conditions of existence from environmental degradation. How can we both negate and defend that which makes our lives possible?

The solution proposed by LEWRG consists of "fighting back oppositionally, rather than simply defending a state we know to be indefensible" (1980, 6). It can be summarised in the following points. First, *In and Against the State* emphasises both the *necessity* and the *insufficiency* of the call to "smash the state."

> The injunction to smash the state is as important now as it ever was. But it is not sufficient. It does not adequately tell the socialist in daily contact with the state what smashing the state means, and how s/he can shape her daily activity in such a way that it becomes part of the struggle for socialism.
>
> *(1980, 86)*

Second, when devising particular anti-state tactics, it is important to recognise the fact that any localised interventions are determined by global historical processes. Consequently, "it is essential that we be aware of what is going on around us, internationally, nationally and in the next department" (1980, 103). Third, public sector workers should acknowledge the contradictory relationship with the state as the point of support for others and source of employment for oneself, which limits the scope of possible anti-state activity ("In choosing how to act to challenge the state we are limited by the hurt we may inflict on other working-class people by doing so. We are limited too, by the fact that we need our jobs" (1980, 103)). Overall, being *in* and *against* the state can be designed as prefigurative struggle, since it involves replacing old social relations with new ones.

Let's translate LEWRG's insights into the language of apocalypticism.

First, we should recognise that the strategy of refusal of the world is *necessary* but *insufficient* as a standpoint. We want the world to end; however, this general orientation borrowed from radical apocalypticism must be supplemented with an analysis of what "the end of the world" means in specific situations and in relation to particular experiences – an analysis characteristic of critical apocalypticism. Such a method – motivated by the refusal of the world, yet attentive to the multiplicity and ambivalence of both reproductive practices and apocalypses – will allow us to develop tactics responsive to a given reproductive context. Here, in line with the intuition underlying critical apocalypticism, we can assert that apocalyptic interventions should address particular forms of reproduction. However, in contrast to critical apocalypticism, the overall goal of such a strategy cannot be the protection of the world, which remains morally and politically indefensible. The radical apocalyptic injunction to end the world remains in force.

Second, any local attacks on specific forms of world's reproduction cannot be separated from the study of apocalyptic tendencies structuring the past, organising the present, and determining the future. By learning about ends which have already happened, or are happening right now, we can anticipate future ends; this anticipation, in turn, can help us to situate ourselves better in relation to previous, existing or coming apocalypses and the worlds they destroy and form. In short, an effective campaign against a particular form of reproduction is located in, and thus depends on, the complex relation of (actual and potential) "ends of the world" and their world-constituting and world-maintaining effects.

Third, earth-bound creatures, including ourselves, depend on the environmental conditions of reproduction. Apocalyptic tactics which undermine our worlds, therefore, must be limited by the damage apocalypses inflict on others (including other species and ecosystems) and on ourselves. Paradoxically, the balance of losses and gains implies that sometimes and for a limited time, worlds may have to be defended. In fact, blocking some ends may be the most effective ways to bring about desired ends (conversely, bringing about certain ends may be the most effective way to block other, undesirable ends). However, from an apocalyptic point of view, tactical support of reproduction can be part of only a wider strategy which favours non-reproduction.

Finally, we should bear in mind that any campaign against the world implies the emergence of new worlds; even though we may struggle against our conditions of existence, in the very act of doing so, we prefigure another world, going against our own strategic injunction to end it. The reader may be reminded here of Mikhail Bakunin's famous quip: "The passion for destruction is a creative passion, too!" (Bakunin 1973, 57).

As it should become clear at this point, being *in* and *against* the world is traversed by multiple contradictions. Apocalypticism expresses a paradoxical standpoint of negating one's own conditions of existence; furthermore, this self-destructive position is limited and undermined by the fact that sometimes, for tactical reasons, it becomes necessary to defend some worlds. Additionally, the avowed attempt to bring about the end of the world always implies a prefiguration of new conditions of reproduction, which introduces with it the risk of endangering and stabilising yet another indefensible world. Non-reproduction, however radical, can never shake off its attachment to reproduction.

Rather than attesting to an essential incoherence and the ultimate redundancy of apocalypticism, the above contradictions make it possible to reconcile apocalyptic orientation with environmental concerns. Apocalypticism can present a needed alternative to dominant ecological strategies motivated by an automatic and undisputed – because ideological – attachment to the reproduction of environmental conditions of existence. If we begin by refusing the world in its entirety on moral and political grounds, that is, if we put into question the attachment to reproduction, we open up a possibility of thinking about the climate crisis *differently*, in opposition to dominant ideologies.

However, because the apocalyptic strategy of refusal is always limited and self-undermining, it can lead to tactical experiments and, perhaps more significantly, to reinventions of planetary forms of existence through its (unwitting) prefigurative power. In other words, by placing itself in and against the world, climate apocalypticism subverts not only the world but also its own apocalypticism; and in so doing, it can generate an ecological position which destroys some worlds, protects others, and engenders new ones. Paradoxically, the hope for the transition to a new, sustainable world may require a passage through a radical negation of current forms of reproduction. As Lynch powerfully puts it, "Let it all go down, so all that suffers its violence or is stifled by its restrictions can live and die differently" (Lynch 2022).

We have arrived at our final hypothesis: perhaps, when all else fails, it is only the end of the world which can save the planet.

A mini-guide for busy readers

Inspired by Jan Sowa's recognition that not everyone will have the time to engage with his work, and the subsequent decision to include a mini-guide for busy readers in one of his monographs (2011), I would like to offer a brief thematic summary which could help readers navigate this book's argument according to their own interests.

The first and second chapters, titled respectively "The answer is ideology!" and "The conceptual orbit of apocalypticism," ask after the *meaning* of climate apocalypse. Chapter 1 aims to account for the polysemy and historical persistence of apocalypticism by uncovering its ideological function of its role in the reproduction of society's conditions of existence. Chapter 2 continues this discussion by reconstructing the semantic field of apocalypticism presupposed by the ideological struggle over the meaning of the end of the world. As I demonstrate, the conceptual boundaries between apocalyptic notions are blurry; as a result, searching for a clear definition of the environmental apocalypse is futile – climate apocalypse is a compound concept which includes (and is included in) other, related terms.

For readers interested in the question of eco-apocalyptic politics, I recommend engaging with Chapters 1, 3, and 7. In Chapter 1, I argue that the role of apocalyptic ideologies is to identify risks to our condition of existence; the paradoxical goal of apocalypticism, therefore, is to ensure the correction of the disastrous threats and the continual reproduction of society. In Chapter 3, titled "Why being is on nobody's side: The politics and ontology of climate apocalypse," I map different expressions of apocalyptic eco-politics, which enables me to question the distinction between authoritarian and preventative apocalypse "from above" and democratic and emancipatory apocalypse "from below." In "Antinomianism and spectral laws," the final chapter of this book, I examine the connection between apocalypticism and a turn to illegal tactics found in Western environmental activism.

The politics of climate apocalypse presupposes specific epistemic and ontological commitments, which I discuss at the end of Chapter 3, as well as in Chapter 4 – "The shapes of eco-apocalyptic time" and Chapter 5 – "A requiem for a world built on sand: Landscapes and the ambivalence of ruins." My argument begins by criticising the belief in a singular end of the world and in the necessity of apocalyptic change, which underlie political expressions of apocalypticism. I then propose an alternative model which acknowledges the temporal multiplicity of apocalyptic ends, as well as their power not only to destroy and create spaces but also to stabilise worlds.

Towards the end of Chapter 5, I argue for the normative ambivalence of apocalypses, which forecloses the temptation to frame the end of the world as a story of redemption. Normativity becomes central in Chapter 6 "When the world ends, I will move to Paris: Anxiety, apathy, and activism." There,

I examine the normative paradoxes of two apocalyptic affects: eco-anxiety and eco-apathy. This discussion opens the question of the normative sources of eco-activism, which I continue in Chapter 7. The final chapter of the book makes a case for the existence of "spectral norms" – a form of normativity aligned to apocalypticism by virtue of its power to challenge the legal foundations of the world. And since spectral norms also found new worlds, they reintroduce the question of the relationship between apocalypticism and the reproduction of our conditions of existence – which should lead the reader back to Chapter 1.

Notes

1 As Lynch acknowledges, his argument finds resonance with antisocial queer theory and afro-pessimism. Cf. Edelman (2004), Wilderson (2020).
2 "[I]f the prophet was a country fellow, it was oxen and cows and so on that were what was represented to him; if he was a solider, generals and armies; and if he was a courtier, a royal throne and such like" (Spinoza 2007, 30).

Bibliography

Althusser, L. (1993) *The Future Lasts a Long Time*. Trans. R. Veasey. London: Chatto & Windus
Althusser, L. (2011) *Philosophy and the Spontaneous Philosophy of the Scientists*. Trans. B. Brewster et al. London: Verso
Althusser, L. (2014) *On the Reproduction of Capitalism: Ideology and Ideological State Apparatuses*. Trans. G.M. Goshgarian. London: Verso
Bakunin, M. (1973) "The Reaction in Germany," in *Bakunin on Anarchy*, S. Dolgoff (Trans. & Ed.). London: George Allen & Unwin Ltd
Danowski, D., Viveiros de Castro, E. (2016) *The Ends of the World*. Trans. R. Nunes. Cambridge: Polity Press
Derrida, J. (1994) *Spectres of Marx: The State of the Debt, the Work of Mourning and the New International*. Trans. P. Kamuf. London: Routledge
Edelman, L. (2004) *No Future: Queer Theory and the Death Drive*. Durham, NC: Duke University Press
Fink, E. (1970) "Appendix VIII: Fink's Appendix on the Problem of the 'Unconscious'," in *The Crisis of European Sciences and Transcendental Phenomenology: An Introduction to Phenomenological Philosophy*, E. Husserl (Ed.); D. Carr (Trans.). Evaston: Northwestern University Press
Levinas, E. (1998) *Of God Who Comes to Mind*. Trans. B. Bergo. Stanford: Stanford University Press
LEWG [The London Edinburgh Weekend Return Group]. (1980) *In and Against the State*. London: Pluto Press
Lynch, T. (2019) *Apocalyptic Political Theology: Hegel, Taubes and Malabou*. London: Bloomsbury
Lynch, T. (2022) "How I Learned to Stop Hoping and Hate the World," Available online: https://cursor.pubpub.org/pub/issue8-lynch-how-i-learned/release/1 [Accessed 19/10/2024]
Possession. (1981) Dir. Andrzej Żuławski. France: Gaumaton International S.A.
Pyne, E. (2023) "Queer Ecologies and Apocalyptic Thinking," in *The Environmental Apocalypse: Interdisciplinary Reflections on the Climate Crisis*, J. Kowalewski (Ed.). Abingdon: Routledge

Secret, T. (2023) "The Improper Apocalypse: Vitalism with and against a Psycho-analytic Approach to the End of the World," in *The Environmental Apocalypse: Interdisciplinary Reflections on the Climate Crisis*, J. Kowalewski (Ed.). Abingdon: Routledge

Sowa, J. (2011) *Fantomowe Ciało Króla: Peryferyjne Zmagania z Nowoczesną Formą*. Kraków: TAiWPN Universitas

Spinoza, B. (2007) *Theological-Political Treatise*. Trans. M. Silverthorne & J. Israel. Cambridge: Cambridge University Press

Taubes, J. (2009) *Occidental Eschatology*. Trans. D. Ratmoko. Stanford: Stanford University Press

Tronti, M. (2019) *Workers and Capital*. Trans. D. Broder. London: Verso

Wilderson III, F.B. (2020) *Afropessimism*. NYC: Liveright

1

THE ANSWER IS IDEOLOGY!

Any mention of *apocalypse* is marked by an equivocation – on the one hand, it can mean a *literal* end of the world (think here of Lars von Trier's *Melancholia* (2011), or Adam McKay's *Don't Look Up* (2021), where earth, and with it the possibility of earthly life, is destroyed); on the other hand, it can be used *figuratively*, to describe a catastrophic event of any size – from a natural disaster or economic breakdown, to a death of a loved one or the experience of redundancy (exemplified by phrases like "my world ended when . . ."). This equivocation, in turn, generates two riddles related to, respectively, the literal and figurative senses of *apocalypse*.

For millennia humanity has been predicting the literal end of the world. Yet the world, indifferent to the prophecies of doom, continues to exist. As history teaches us, the ongoing disappointment which follows apocalypticism hasn't been able to deter new prophets from appearing. The fact that previous generations wrongly believed that the end was nigh makes no difference to *today's* prophecy. There is something about apocalypticism which has a power to bury its past failures, and to reappear as a real possibility in the present, disconnected from its previous ill-fated instantiations.

Here is our first question: why is apocalypticism so persistent, despite thousands of years of disappointment? This problem is particularly relevant today, when the environmental crisis seems to warrant a serious engagement with the notion of *climate apocalypse*. It remains a puzzle why apocalypticism, continually discredited by the existence of the world, imposes itself again and again as tempting frame, seductive enough to spill some ink to argue for or against it.

Admittedly, it can be argued that contemporary environmental apocalypticism operates with the figurative notion of the apocalypse. However, the

DOI: 10.4324/9781003348511-2

figurative reading makes it difficult to identify what the term actually means: it can be a crisis, a catastrophe, or an extinction event; it can have unambiguously negative connotations, or, on the contrary, it can evoke the images of reckoning and redemption. Importantly, however, this polysemy of meanings doesn't prevent the notion of apocalypse from being used with ease, and in a seemingly self-explanatory way, in mainstream Western debates on ecology.

Here is our second question: why is *apocalypse* immediately understood, despite apparently possessing neither a stable meaning nor fixed referent?

These two riddles, historical and semantic, are in fact related. The thesis which I will advance in this chapter is that both the literal and the figurative use of "the end of the world" are expressions of *ideology*. More specifically, apocalypticism communicates, in a distorted way, the concern with the reproduction of our conditions of existence characteristic of ideology. The ideological character of "the end of the world" across its different instantiations, in turn, enables us to explain both the historical persistence of apocalypticism, despite its long-term failure, and the sematic straightforwardness with which apocalypse is employed and received, despite the bewildering polysemy of the term.

Even though in this chapter I will focus my analysis on apocalypticism *in general* – which requires a "bracketing" of particular instantiations of apocalypticism – the conclusions will be of direct relevance to the specific case of *environmental* apocalypse. This is because approaching "the end of the world" from the point of view of ideology will generate a philosophical frame capable of making sense of the complexities which plague both the contested meaning of climate of apocalypse and the messy politics of eco-apocalypticism. This chapter, therefore, offers a *general* theory of apocalypticism, contributing the philosophical foundations necessary to develop a *special* theory of eco-apocalypse in the following chapters.

The importance of failure

Let's begin by considering two interpretations of apocalypticism. On one reading, apocalypticism can be characterised as a subjective predisposition – a lens through which we experience and interpret the world. Such a lens, after its initial appearance in history, can be passed across generations, either by being internalised by individual subjects or by being maintained by collectives.

But the exclusively "subjectivist" reading can't account for the fact that apocalypticism resurfaces, *each time anew*, in response to particular worldly events. Eco-apocalypticism is a case in point: its relevance is better explained by the ongoing extreme weather events – that is *specific objective factors* – than by inherited subjective optics.

In turn, a critic of the "objectivist" position may reply that literal apocalypse is always an imagined object. What we can point to in the world are only parts of reality which can be called apocalyptic (e.g. catastrophic climate change); however, these worldly objects are always *signs* of the apocalypse, which is located in an always deferred future. Any "end of the world" is only an imaginary referent which helps to unify disparate disastrous events – a claim further supported by the continuous existence of the world.

Interestingly, historical examples of apocalypticism suggest it consists of an interaction between subjective and objective poles; consequently, apocalyptic prophecy shouldn't be reduced solely to either subjective predispositions or objective factors. The interaction between subjects and objects becomes particularly visible in the response of apocalyptic movements to failed eschatological prophecies.

As the authors of *When Prophecy Fails* observe, when their predictions are disappointed and the world continues to exist, apocalyptic groups don't abandon apocalypticism, as we may anticipate; in fact, in the aftermath of a prediction that didn't come to pass, we witness no change to their commitment to "the end of the world." Certainly, members of apocalyptic movements experience a psychological dissonance, but they are able to reduce it by mustering social support for their belief through an increased proselytising. The larger the number of fellow believers, the lesser the psychological dissonance between belief and reality (Festinger et al. 2008). The search for new followers is often accompanied by a revision of doctrine. As Jacob Taubes notes, the non-arrival of the end of the world is often integrated into an updated and more complex theoretical system, which nonetheless continues to maintain the overall apocalyptic orientation (Taubes 2004). Armed with social support and a new doctrine, apocalyptic groups then embark on the search for new signs capable of confirming the revised theses of their apocalypticism (Festinger et al. 2008). Apocalypticism, therefore, persists *in spite of its failure*, thanks to social support and a keen eye for apocalyptic signs, and, *because of its failure*, when the disappointed expectation becomes an integral feature of the wider eschatological doctrine.

We can identify a circular process at play between subjective and objective poles: first, the end of the world is prophesied; second, the prophecy is proven wrong by the continual existence of the world; third, corrected prophecy of the end is proposed – which itself will most likely be disconfirmed by the survival of world.

The strategies aimed to reduce the sense of failure contribute to the polysemy of the term "apocalypse." Every time prophecy is disappointed, the meaning of the awaited end must be reformulated: in addition to making sense of the non-arrival of the initial apocalypse, it must resonate with both the expanded support base and the new signs. The multiple meanings of apocalypse could, therefore, be (in part) explained as a historical testament to the creative confrontation with the failure of apocalypticism.

While failure can illustrate how the subjective and objective poles interact within apocalypticism, this process doesn't really seem like the whole story. Think here of everyday life: we make plans which end in failure on a regular basis; we then adjust our expectations to mitigate the initial disappointment. But if we fail *every single time*, we finally give up – unless, of course, we are *compelled* to repeat the failure over and over again by some deep-seated and often unconscious causes. What I will argue here is that apocalypticism, in analogy with subjective compulsion to repeat, is also animated by deep-seated and unconscious reasons which explain its persistence and longevity despite the never-ending disappointment. To explain these causes, we must turn to the analysis of ideology.

Apocalypticism and ideology

In this section, drawing on the work of Louis Althusser, I would like to foreground the ideological character of apocalypticism.

In everyday language, we refer to ideology as a set of ideas (political, religious etc.) which, at best, colour, and at worst, completely distort our experience of the world. In the rather technical words of Louis Althusser, the everyday concept of ideology expresses the fact that people "represent their real conditions of existence to themselves in an imaginary form" (2014, 256). But, Althusser tells us, the everyday definition of ideology must be corrected:

> [I]t is not their real condition of existence, their real world, that "men" "represented to themselves" in ideology, but above all it is their relation to those conditions of existence which is represented to them there . . . it is the *imaginary nature of this relation* which underlies all the imaginary distortion that we can observe.
>
> *(2014, 257)*

For Althusser, ideology is primarily not a set of ideas which block our access to the world as it really is. Rather, ideology is first and foremost an imaginary representation of our *relationship* with the real world. It is the imaginary character of the relationship between me and the world that is responsible for any other biases and prejudices which we normally associate with ideology. To put it simply, how I view object X depends on my relation to object X; consequently, if my relation to object X is distorted, this may result in a distorted view of object X. A good example here is the belief in fate: if I experience my relationships with the world as structured by providence, then particular events will seem to me as part of a wider plan of gods or the universe.

While ideology constitutes "a system of representations," it doesn't impose itself directly on our consciousness; instead, Althusser points out, the mediation between the subject and the object ideology achieves happens through structures (e.g. apparatuses and practices) whose functioning, for the most part, escape us. Ideology, therefore, is analogous to imagination in Spinoza – rather than being simply an internal operation of the mind, it is the effect of the way in which external bodies (or structures) affect our bodies, generating ideas of things (Spinoza 1996). Importantly, because for the most part we are not aware that we are *in* ideology, we mistake ideological mediation for an immediate experience of the world. As Althusser puts it, people " 'live' their ideologies as the Cartesian 'saw'. . . the moon two hundred paces away . . . as their '*world*' itself" (Althusser 1969, 233).

One of the consequences of "living in" ideology is that ideological representations, while causing our experiences and beliefs, are frequently seen as the *effects* of our seemingly immediate interactions with the world. To return to our example of fate, oftentimes the belief in higher power governing the universe appears as an *outcome* of a series of providential experiences ("I can't believe x,y,z happened to me; the universe must be looking out for me"). In reality, the series of events are experienced as providential *because* they are mediated by the imaginary representation of divine forces ruling the universe.

To complicate matters a bit, we should also note that ideological representations which cause an experience may *distort* the representations which are the outcome of the experience. To stick to the example of fate, we can say that the *conscious* beliefs in gods or the will of the universe are in fact misrepresentations of the lack of control I *unconsciously* experience over my life. It is the latter which mediates between me and the world; consequently, it is the lack of control which results in the experience of the world as providential and the subsequent conclusion that these events are part of a divine or cosmic plan. In other words, fate *distort*s and *hides* the experience of uncontrollable reality by endowing occurrences with providential significance and connecting them with a plan of a higher power.

To sum up, ideology is a representational structure which mediates between subject and the world and which hides itself by misrepresenting this very relation. This definition, in turn, dovetails with the description of apocalypticism I proposed in the previous section. We can reformulate the *interaction* between subjective and objective poles which characterises apocalypticism as a *mediation* between subject and the world. This means that apocalypticism is primarily not a representation of the ending world but a representation of a *relation* to the world, which makes the latter appear as ending. Furthermore, if we follow Althusser, we can say that apocalypticism *misrepresents* this relation. On the one hand, we have already observed a surface-level distortion: after apocalyptic prophecy fails, the existence of the world is incorporated into a new apocalyptic belief system – the survival of the world, rather than contradicting apocalypticism, becomes an integral part of the latter. On the other hand, we can expect to encounter more profound distortion: apocalypticism would misrepresent and dissimulate its ideological effects. To examine the latter distortion in more detail, we must discuss the twofold function of ideology.

Ideological representations, by mediating between subjects and their environment, enable us to effectively navigate, use, and transform the world around us. More precisely, structures, apparatuses, and practices – by affecting our bodies and by producing images of things – provide us with the know-how needed to manipulate reality. Ideology, therefore, "is indispensable in any society if men are to be . . . equipped to respond to the demand of their conditions of existence" (Althusser 1969, 235). However, the know-how generated by ideology has a specific function: its goal is to ensure that the way we relate to our condition of existence *reproduces* these very conditions (Althusser 2014). In other words, living in ideology should guarantee that our environment and the life it makes possible continue to exist.

I would like to suggest the following hypothesis: just like other ideologies, apocalypticism helps us to navigate the world, and, in so doing, it expresses the demand for the reproduction of our conditions of existence. What accounts for the specificity of apocalypticism is its paradoxical structure, which distinguishes it from other ideologies: the demand for reproduction is articulated in and as the image of "the end of the world." More precisely, the goals of apocalyptic prophecy are to identify *threats* to reproduction and eliminate the danger through a re-optimisation of our practices. To put it metaphorically, apocalypticism identifies symptoms of a potentially terminal illness, which, in turn, allows the application of the cure and the survival of the world. This ideological function, therefore, radicalises the significance of failure for apocalypticism: disappointment is not only the reason for internal changes to apocalyptic doctrine and expectation; ensuring the failure of apocalyptic prophecy is also the ideological reason for prophesying the end – if the function of images of the end is to reproduce our conditions of existence, then the ideological goal of even the most expectant apocalyptic prophecy is to prove itself wrong.

As Jean-Pierre Dupuy has shown in his account of "enlightenment doomsaying," prophecies often admonish the sinful by giving them a choice to either repent or have their world destroyed. A good example here is Jonah, who reluctantly warns the people of Nineveh of their imminent destruction and who "knows that his prophecy, in acting upon the world – *by virtue of the very fact of its acting upon the world* – will become false" (Dupuy 2022, 111). While I like Dupuy's reading of Jonah, my claim is more extensive: my contention is that apocalypticisms which aim to *bring about* the end of the world are equally expressive of the demand for the reproduction of our conditions of existence. Despite their conscious intentions, the excited expectations of apocalypse unconsciously play a role in *avoiding* the end of reproduction. In contrast to Jonah, these prophets of doom *don't know* that their prophecy will become false because of its presence in the world.

Here we encounter the distortion of cause and effect proper to apocalypticism *qua* ideology. While we may think that reality is populated with catastrophic sign because of the coming apocalypse, this conscious expectation is, in fact, caused by the unconscious demand for the reproduction of our conditions of existence proper to ideology.

The historical riddle

The ideological function of apocalypticism can help us solve the historical puzzle of the persistence of apocalypticism despite its consistent failure.

Althusser distinguishes between particular ideologies with determinate histories and ideology *in general*, which, as he puts it, "has no history":

> [T]he peculiarity of ideology is that it is endowed with a structure and a functioning such as to make it a non-historical reality, i.e. an *omni-historical* reality, in the sense in which that structure and functioning are immutable, present in the same form throughout what we can call history . . . our proposition – ideology has no history – can and must . . . be related directly to Frued's proposition that the *unconscious is eternal*, i.e. that is has no history. If eternal means, not transcendent to all (temporal) history, but omni-present, trans-historical and therefore immutable in form through the extent of history.
>
> *(2014, 255)*

We have already seen why ideology in general can function as an omni-historical element: its role is to ensure that people can relate to their world in a way which secures the reproduction of their conditions of existence. Insofar as societies must reproduce their environment to survive, ideology "is not an aberration or a contingent excrescence of History: it is a structure essential to the historical life of societies" (Althusser 1969, 232).

As I have proposed, the ideological function of apocalypticism is to identify and facilitate the response to the threats to reproduction. If we think of ideology as a machine, apocalypticism would be the red light which flashes whenever the machine overheats or when its parts need to be replaced. But is apocalypticism an aspect of ideology *in general* or is it one of many particular ideologies? Is apocalypticism an inbuilt, factory setting of the ideological machine, or is it, on the contrary, only a contingent side effect of particular ideological circumstances?

I believe that the widespread persistence of apocalypticism can be accounted for, and our historical riddle can be answered, if we posit apocalypticism as an essential aspect of ideology in general – that is to say, if we propose that apocalypticism, by virtue of its ideological function, is "a structure essential to the historical life of societies." Of course, I don't mean suggest that every culture speaks of apocalypse in the same way: its concept has a determinate history and particular expressions of apocalypticism (e.g. ancient Jewish apocalypses, Cold War apocalypse) all reflect specific historical environments. My point is, rather, that whenever ideology is at work, apocalypticism can be found too: there will always be a mechanism for identifying, amplifying, and (as I will discuss later) exaggerating dangers to reproduction, since such a mechanism is necessary for the effective survival of society. In the Althusserian sense, apocalypticism in general *has no history*.

This conclusion enables us to explain the longevity of apocalypticism: its reappearance throughout history is an expression of its omni-historical ideological function; more importantly, however, its appeal is dependent *not* on people believing in the imminent apocalypse (such a belief is constantly disconfirmed by the existence of the world) but on the representational structures mediating between people and their conditions of life, which impose themselves on subjects in every society. Consequently, apocalypticism is indifferent to the successes of previous reiterations of apocalyptic expectation, because the "force" behind it is generated not by the imminent end of the world but by the trans-historical demand for the reproduction of our conditions of existence inherent in ideology. It is the ideological function of "the end of the world" which operates as the deep-seated and often unconscious cause of the compulsion to repeat apocalypticism, both *in spite of* and *because of* the ultimate failure of the apocalyptic prophecy.

The semantic riddle

The intimate relationship between ideology and apocalypticism can also explain how the term apocalypse can be used meaningfully and in a straightforward manner, despite its rather confusing polysemy.

As Althusser notes, ideology is a representational structure which involves images and concepts (Althusser 1969, 233). More specifically, images and concepts can *summarise* our relationship to the world mediated by ideology. For example, "the bourgeoisie *lives* in the ideology of *freedom* the relation between it and its conditions existence" (Althusser 1969, 234). Here the notion of "freedom" functions as an abbreviation of a more complex ideological structure in which the bourgeoise find themselves.[1]

This means that ideological concepts signify two things at once: first, the notion of "freedom" can *refer* to concrete instantiations of freedoms found in a bourgeois society – the freedom of speech or the freedom of assembly.

On the other hand, insofar as "freedom" summarises the bourgeois ideology, the *sense* of the notion, that of which it is expressive, is its ideology. We can be more specific here – the abbreviating concept would express (albeit in a distorted and often unconscious way) the ideological demand for reproduction of specific conditions of existence – in the case of "freedom," the conditions of bourgeois society.

The notion of apocalypse plays a role analogous to "freedom": it abbreviates the complex representational structure within which it operates. Apocalypse signifies, on the one hand, the concrete instantiations of "the end of the world" (what we can call its *referents*), and on the other hand, the ideological demand for the reproduction of specific conditions of existence (which we can call its *sense*). Now, it is the ideological sense which determines which referents can be picked up by the summarising concepts. When the relation between a concept and an object is obvious, this is primary because the object has been "pre-appointed" to be recognised as an obvious instance of the concept by ideology:

> It is indeed a peculiarity of ideology that it imposes (without appearing to do so, since these are "obviousnesses") obviousness as obviousness, which we cannot *fail to recognize* and before which we have the inevitable and natural reaction of crying out (aloud or in the "silence of consciousness"): That's obvious! That's right! That's true!
>
> *(Althusser 2014, 262–263)*

Let's use Althusser's own example to illustrate this point: the unborn child. Even in early stages of pregnancy, we already know that it will bear the family name, it will be assigned sexual identity, it will be described as looking like the parents etc. "Before its birth, the child is . . . appointed as a subject in and by the specific familial ideological configuration in which it is 'expected' once it has been conceived" (Althusser 2014, 265). When, after its birth, the child does indeed bear the family name, has a sexual identity, and seems to resemble its parents, this fulfilment of the pre-assigned expectation is recognised as *obvious* – of course the child would possess all of these characteristics, since this what it means to be a child!

Analogous "rituals of recognition" take place in the expectation of the end of the world (I will explore the analogy between birth and the apocalypse in more detail in Chapter 5). Even before they occur, some practices, events, and processes – specifically those which indicate a danger to reproduction – are "pre-appointed" as apocalyptic, and, therefore, immediately recognised as such when they do take place. Apocalypticism, therefore, offers an "ideological configuration" in which any threat to reproduction, however diverse, is recognised as an obvious sign of the apocalypse. Consequently, the term apocalypse has no trouble in picking out its often heterogenous referents in a straightforward way, since the latter have been pre-appointed as obviously related to "the end of the world."

The rituals of recognition, therefore, account for the polysemy of apocalypse: the multiple meanings of "the end of the world" (in addition to bearing the marks of prophets' creative confrontation with the failure of their prophecy) are generated by that fact that multifarious practices, events, and processes have been pre-assigned an apocalyptic status by virtue of constituting potential risks to reproduction. To put it in linguistic terms, the referents of "apocalypse" can be different because its ideological sense remains the same. This is particularly visible in the infamous equivocation between the literal and figurative meanings of the "the end of the world" which haunts apocalypticism. While their referents are clearly heterogeneous – one designating a threat *to* the world, the other referring to catastrophes *within* the world – the literal and figurative apocalypses share the same sense: they both pick out risks to the reproduction of our conditions of existence.

We can propose two further hypotheses regarding the relationship between literal and figurative "ends of the world": first, we can expect to encounter more threats to reproduction *within* the world and then threats *to* the world. While the earth may be destroyed by a meteor, the more common risks to reproduction are specific, inner-worldly practices, events, and processes. Consequently, the figurative meaning of apocalypse would be used more frequently than its counterpart. Second, and following from the previous point, if the figurative meaning of apocalypse is more widespread, then for the most part the talk of the end of the world would be *exaggerated* – the apocalypse would simply be an inflated image of smaller-scale risks. Interestingly, the same conclusions have been arrived at by the medieval philosopher Moses Maimonides.

A brief detour: Maimonides and "the end of the world"

For Maimonides, the prophets communicate their visions by means of "figures, hyperboles, and exaggerations" – in fact, prophetic words

> would create strange ideas if we were to take them literally without noticing the exaggeration which they contain, or if we were to understand them in accordance with the original meaning of the terms, ignoring the fact that these are used figuratively.
>
> *(Maimonides 1956, 247)*

The use of figurative language is in part related to the fact that the visions themselves are frequently experienced as allegorical representations: images ("such as the candlesticks, horses, and mountains of Zechariah" or "the scroll of Ezekiel") stand for other ideas, which are often indicated by means of linguistic relations: for example, "the imaginative faculty forms the image of a thing, the name of which has two meanings, one of which denotes something different [from the image]" (Maimonides 1956, 239).

For reasons which I will explore in further chapters, Maimonides brackets the question of the *literal* end of the world – for him, one can reasonably embrace either the eternity or the finitude of the universe. The significance of apocalyptic prophecy, Maimonides tells us, lies elsewhere: in *The Guide for the Perplexed*, Maimonides proposes to read apocalyptic prophecies as hyperboles with a strictly political meaning. On Maimonides's interpretation, "the end of the world" is a figurative and exaggerated image which can represent "the ruin of a kingdom or a destruction of a great nation" (1956, 204). Commenting on apocalyptic verses of Isaiah 34:4, Maimonides writes:

> Will any person who has eyes to see find in these verses . . . anything but a figurative description of the ruin of the Edomites, the withdrawal of God's protection from them, their decline, and the sudden and rapid fall of their nobles? The prophet means to say that the individuals, who were like stars as regards their permanent, high, and undisturbed position, will quickly come down, as a leaf falleth from the vine, and as a fig falling from the figtree.
>
> *(1956, 206)*

Interestingly, Maimonides also notes that prophecy can employ apocalyptic imagery to speak of the re-establishment of a kingdom (sometimes as a result of a fall of another realm). For example, when Isaiah says "I will create new heavens and a new earth. The former things will not be remembered" (Isaiah 65:17), the prophet, in fact, refers to a restoration, stability, and permanence of a kingdom "described as a creation of heaven and earth" (Maimonides 1956, 207).

It is striking that Maimonides, in his own idiom, echoes the conclusions we reached on a basis of the ideological reading of apocalypticism. The images and words of apocalyptic prophecies have a twofold signification: they have an immediately accessible surface level meaning, and a more profound, allegorical meaning in need of articulation. The allegorical images of "the end of the world," in turn, have a political sense: they are figurative representations of ends of specific societies, coupled with exaggerated pictures of improved forms of collective life. In short, already for Maimonides, apocalypticism is a set of imaginary representations which both distorts and bears directly on the political questions related to reproduction of the conditions of existence of given societies.

Importantly, for our purposes, the example of Maimonides's proto-ideological interpretation of apocalypticism demonstrates that the problem of ideology is not an arbitrary or external imposition, which forces the apocalyptic tradition into an alien theoretical frame. On the contrary, as Maimonides powerfully shows, the questions raised by an analysis of ideology can also be posed from *within* traditional apocalypticism. In other words, while we can apply ideology to apocalypticism (as I have done throughout most of this chapter), it is also possible to arrive at ideology starting with the apocalyptic tradition – even in its most orthodox form of Biblical prophecy.

The subject, the world, and the struggle

To conclude this chapter, I would like to briefly sketch three further sources of the polysemy of apocalypse: subjective inflections, the ambivalence of the world, and ideological struggle.

The Althusserian reading of apocalypticism may give the impression that subjective experience is solely an *effect* of ideology. A Maimonidean supplement, which I believe is of central importance, lies in recognising that apocalypticism must be expressed and that this expression leaves a personal mark on the message communicated: "every prophet has his own peculiar diction, which is, as it were, his language, and it is in that language that the prophecy addressed to him is communicated to those who understand it" (1958, 204). In other words, apocalypticism must pass through the subject; this expression in turn "personalises" the end of the world. We can say that prophetic subjectivity, as well as being an effect of ideology, is necessary to *translate* the unconscious ideological demand for reproduction into a conscious message. Consequently, to use an Althusserian turn of phrase, it is the subjective practice of prophecy which ensures the material existence of apocalypticism. This subjective work of prophetic translation, in turn, impresses on the apocalypse the specific characteristics of the prophet. Here we encounter another reason for the polysemy of apocalypse: in addition to a response to failure and rituals of recognition, each prophecy alters the meaning of the "end of the world" in accordance with the prophets' language and "peculiar diction." As I will show in the later parts of this book, the necessary passage of apocalypticism – including its environmental instantiation – through subjectivity, produces a wide range of experiences which extend well beyond prophecy. The analysis of these experiences is significant for the reasons implied by Maimonides: to understand the modulations of apocalypticism, we must pay attention to how "the end of the world" is expressed in the personal idiom of concrete subjects.

The polysemy of "the end of the world" can also stem from the fact that the meaning of "the world" is *ambiguous*, since it can meaningfully refer to individual, collective, and planetary environments. Furthermore, the world is experienced as *ambivalent*: on the one hand, it is a condition of life, a source of nourishment and joy; on the other hand, it presents incessant obstacles to flourishing existence – so much so that some have rejected the concept of "the world" altogether on ethico-political grounds (Lynch 2024). The examples of the ambiguity and the ambivalence of the world can be found in eco-apocalyptic discourses. The discussions of the environmentally induced world collapse are often haunted by the lack of clarity on the questions of *whose* world may end, as well as *whose* world is being protected. Furthermore, the calls to tackle the climate crisis by ending the unjust world are often compromised by the awareness of the costs for the most vulnerable associated with adapting to a post-apocalyptic situation. I will return to the difficulties generated by the concept of the world in Chapter 5.

The final source of polysemy of apocalypticism, operating alongside the subject and the world, can be uncovered by passing through the dream-like character of prophecy. As Maimonides notes, the difference between prophecy and dreams is "one of quantity, not of quality": "our Sages say, that dream is the sixtieth part of prophecy" (1956, 225). Dreams, of course, are royal road to the unconscious (Freud 1997) – the part of psychic life which Althusser compares to ideology (2014, 255). Loosely associating dreams, prophecy, the unconscious, and ideology, we can suggest the following analogy: if ideology is like the unconscious, then dream-like prophecy is equivalent to the conscious processes, which express, in the characteristically distorted way, the unconscious level of ideology.

However, what is most interesting about this topography is what it *misses*: Freud also identifies a *preconscious* level located in between the conscious and the unconscious processes (Freud 1981, 190–194). This poses the question: what, in our analogy, plays the role of the preconscious level found in between prophetic subjectivity and ideology? The answer I would like to propose is – *ideological struggle*.

In a recent study, G. Anthony Keddie suggests that ancient apocalypses are in fact competing ideologies, "political refractions of reality with the objective of self-legitimation" (2018, 7). Keddie takes issue with interpretations of apocalypses which view the genre through an emancipatory lens – a reading which considers apocalyptic literature as proto-revolutionary manifestos. Instead, Keddie argues, apocalypses constitute:

> the attempt to displace a dominant ideology with an alternative ideology that affirms most or all of the same structural foundations of social and economic inequality while legitimating a different set of political authorities. In the shifting structures of economic and cultural change, local contingents of elite or subelite scribes objectified causes of inequality and denounced other provincial elites as responsible for them. These texts contributed to social transformation by generating class dispositions that ideologically refracted actual socioeconomic structures of inequality for political legitimation.
>
> *(2018, 6)*

The recognition of apocalypticism as a site of political antagonism was already present in Maimonides's *The Guide*, where the prophesied restoration of a kingdom presupposed the collapse of a competing realm. Kiddie's approach enables to further develop this insight by making explicit the dimension of *ideological* struggle inherent to apocalypticism.

Particular apocalyptic prophecies are traversed by two intersecting lines of antagonism: there is a struggle against the dominant ideology and an opposition to other instantiations of apocalypticism, especially (although not exclusively) when the latter become dominant ideologies. Although the struggle against dominant ideology (apocalyptic or not) lends itself to a romantic image of an underdog standing up to the oppressors, we should bear in mind Kidde's warning: while apocalypticism can enact social transformation, apocalyptic ideology is not *necessarily* emancipatory or revolutionary. There can exist reactionary, conservative, and authoritarian instances of apocalypticism, where the change effected by apocalyptic rhetoric serves not to liberate the masses but to legitimise new power structures. Ideological struggle, therefore, offers another cause of the polysemy of "the end of the world." Part of the ideological struggle involves a conflict over the correct meaning of apocalypse, whose semantic content becomes blurry when the clash is multilateral, and each side operates with its own "end." For example, when a political group uses the specific images of the apocalypse to attack its opponents, the latter are confronted with their own "end of the world": the catastrophic possibility of the loss of power.

In the following chapters, I will initially focus on the "pre-conscious" level of eco-apocalypse: I will approach environmental apocalypticism as a site of political antagonism, expressed in part in the struggle over the meaning of climate apocalypse. This analysis, in turn, will lead me to reintroduce, albeit from a different angle and among other themes, the two other sources of (eco-)apocalyptic polysemy: subjective experiences and the ambivalent character of the world.

Note

1 Similarly, the changes in communist ideology of USSR after the death of Stalin can be summarised by the term "humanism." "Soviet Union has proclaimed the slogan: All for Man, and introduced new themes: the freedom of the individual, respect for legality, the dignity of the person" (Althusser 1969, 221).

Bibliography

Althusser, L. (1969) *For Marx*. Trans. B. Brewster. Harmondsworth: Penguin Books
Althusser, L. (2014) *On the Reproduction of Capitalism: Ideology and Ideological State Apparatuses*. Trans. G.M. Goshgarian. London: Verso
Don't Look Up. (2021) Dir. Adam McKay. USA: Hyperobject Industries

Dupuy, J.-P. (2022) *How to Think About Catastrophe: Toward a Theory of Enlightened Doomsaying.* Trans. M.B. DeBevoise & M.R. Anspach. East Lansing: Michigan State University Press

Festinger, L., Riecken, H.W., Schachter, S. (2008) *When Prophecy Fails: A Social and Psychological Study of a Modern Group that Predicted the Destruction of the World.* London: Pinter & Martin Ltd

Freud, S. (1981) "The Unconscious," in *The Standard Edition of the Complete Psychological Works of Sigmund Freud – Vol. XIV.* Trans. J. Strachey. London: The Hogarth Press and the Institute of Psycho-Analysis

Freud, S. (1997) *The Interpretation of Dreams.* Trans. A.A. Brill. Ware: Wordsworth Editions Limited

The Holy Bible. (2011) *New International Version.* Palmer Lake: Biblica

Keddie, G.A. (2018) *Revelations of Ideology: Apocalyptic Class Politics in Early Roman Palestine.* Leiden: Brill

Lynch, T. (2024) "A Political Theology of the World that Ends," in *Worlds Ending. Ending Worlds,* J. Stümer & M. Dunn (Eds.). Oldenbourg: De Gruyter

Maimonides, M. (1956) *The Guide for the Preplexed.* Trans. M. Friedländer. New York: Dover Publications, Inc.

Melancholia. (2011) Dir. Lars von Trier. Denmark: Zentropa

Spinoza, B. (1996) *Ethics.* Trans. E. Curley. London: Penguin Books

Taubes, J. (2004) *The Political Theology of Paul.* Trans. D. Hollander. Stanford: Stanford University Press

2

THE CONCEPTUAL ORBIT OF APOCALYPTICISM

What does "climate apocalypse" mean? This chapter aims to answer this question by situating the notion of eco-apocalypse within the conceptual history of apocalypticism. More specifically, my goal is to show that the notion of the environmental apocalypse is richer than we may initially expect, since it contains semantic aspects of related terms (such as catastrophe, extinction, revolution, or war, among others). Overall, my discussion demonstrates that the semantic field of apocalypticism is essentially *blurry*. Apocalyptic notions internalise or "swallow-up" other terms, turning themselves into compound concepts with multiple meanings; in so doing, they erase any clear conceptual boundaries between themselves and their counterparts. Climate apocalypse is a case in point: as I will show, it subsumes semantic elements found, for example, in religious and nuclear apocalypses.

Certainly, a critic may point out, for instance, that when they use the term "environmental crisis," they specifically don't mean climate apocalypse; conversely, when they bring up "the end of the world," they do so because they find the notion of crisis inadequate. Furthermore, the critic could add, blurring apocalyptic meanings flies in the face of respectable philosophical practice, which consist of drawing – and not collapsing – conceptual distinctions. This chapter's approach, therefore, would be equivalent to resigning ourselves to an unacceptable non- or pre-philosophical standpoint – not to mention that I will fail to answer what climate apocalypse means.

While I sympathise with the critic, I nonetheless will attempt to show (to the horror of philosophers) that the semantic boundaries between apocalyptic concepts are indeed blurred and in constant mutation and that, consequently, (to the dislike of speakers priding themselves on semantic precision)

DOI: 10.4324/9781003348511-3

in employing an apocalyptic term such as climate apocalypse we always mean more than we intend.

In this chapter, therefore, I will attempt to lead the reader through the labyrinthian connections established between concepts found in the semantic orbit of eco-apocalypticism. While for some readers it may appear as if I am consciously obscuring the discussion of climate apocalypse by confusing its definition, my aim is, rather, to show the semantic complexity of the concept of environmental apocalypse and its concomitant relationships with diverse theoretical and historical contexts. I conclude this chapter by proposing that the semantic elements which circulate in the conceptual orbit of apocalypticism are "gathered together" by ideology.

Before beginning the discussion, I would like to make two additional qualifications. First, my goal is not to present a chronological accumulation of apocalyptic meanings across history – even if at times I do follow linear time. Rather, I hope to construct semantic constellations which cut across chronology; this chapter's approach is perhaps comparable to a science fiction story in which the protagonists "jumps" between theoretical and historical worlds. Second, while the ecological end of the world can stand for a planetary or even a cosmic theme, the apocalyptic notions which designate it reflect specific geographical, historical, and theoretical conditions of their emergence. This is also true of my analysis – I present only one of many possible semantic constellations constitutive of the orbit of apocalypticism, which is undoubtedly fragmentary and which deals mainly with texts and events which belong to the broadly construed Western tradition.

Apocalypse with and without kingdom

One of the most prevalent ways of differentiating apocalypses is to categorise them as either *redemptive* or *destructive* "ends of the world." Günther Anders, for example, distinguishes between apocalypses which usher a new world or a "kingdom" – paradigmatically exemplified by the Judeo-Christian eschatology – and "naked apocalypse without kingdom." The latter, typified by nuclear apocalypse, is defined in opposition to its religious counterpart as a pure annihilation, a secular event completely devoid of any redemptive elements (Anders 2019).

At first glance, climate apocalypse would fall into the category of apocalypse without kingdom: it would stand for a secular and catastrophic end of the world, which, as such, should be averted. To believe that the environmental end of the world should be awaited, or that it possesses some type of redemptive or even religious significance, would be utterly irresponsible.

However, in his article, Anders splits the notion of the kingdom – or of a redemptive future – into its religious and secular expressions, which begins to blur distinction between the two types of apocalypses he initially identified.

For Anders, revolutions can be labelled as secular apocalypses *with* a kingdom ("The schemas of Judeo-Christian eschatology . . . shone very clearly through the Communist doctrine . . . Marx and Paul seem to become contemporaries" (2019)). Here, in analogy with religious apocalypses, history proceeds towards a final event of history capable of redeeming the suffering of the oppressed classes. But, similarly to the nuclear apocalypse, the responsibility for the fulfilment of history is placed not in the hands of God but in those of humanity, who become the agents of the apocalyptic event.

Importantly, for our purposes, the introduction of revolution as an exemplar of apocalypse with a kingdom complicates the seemingly straightforward categorisation of the environmental end of the world as apocalypse *without* kingdom. While the catastrophic character of eco-apocalypse undoubtedly places it in close proximity to its nuclear counterpart, any mention of large-scale ecological transformations which often accompanies the analysis of climate disasters begins to formulate a vision of a kingdom and the steps needed to get there. In other words, the possibility of a kingdom brought about by the action of humanity, characteristic of revolution, shines through – however feebly – whenever the discussion of eco-disasters turns to proposals for a more just and sustainable future.

Admittedly, one could argue that climate apocalypse with a *secular* kingdom would remain radically distinct from its *religious* counterpart. However, as I will show in the next section, Biblical ends of the world are much closer to secular apocalypses than we may initially suspect. Although Judeo-Christian apocalyptic literature ascribes agency to God, it nonetheless endows humanity with responsibility for both the unravelling and the outcome of the apocalypse. Furthermore, both religious and secular apocalypses – for example the Final Judgement and nuclear catastrophe – include environmental events as part of their narratives. This means, on the one hand, that Biblical apocalypses are not *straightforwardly* redemptive, since they include also purely catastrophic occurrences – including climate disasters (as such, Judeo-Christian ends of the world would begin to blur the lines between the apocalyptic labels proposed by Anders); and, on the other hand, that climate apocalypse is an integral element of apocalypses regardless of whether they are found in the Bible or in secular discourses or whether they operate with or without the idea of a kingdom.

Biblical apocalypses and environmental ends

In the context of climate apocalypse, humanity is a key actor who brings about, exacerbates, and has the power to address the catastrophe. By contrast, in Biblical apocalypses – at least according to the thesis put forward by Gershom Scholem and Jacob Taubes in their respective works – humanity is presented as devoid of agency. "Can man master his own future? And the answer of the apocalyptist would be: no" – writes Scholem (1971, 15). Taubes echoes this sentiment: "The Kingdom of God is to become a reality without having to pass over the threshold of the human will" (2009, 34). The divergent answers to the question of human agency, in turn, would constitute an unbridgeable divide between Judeo-Christian and ecological apocalypse, rendering Biblical apocalypticism incompatible with its environmental counterpart.

However, it is my contention that the interpretation of agency in Judeo-Christian apocalypticism suggested by Scholem and Taubes is exaggerated. It is clear that Biblical apocalyptic literature acknowledges the importance of human action. In other words, even if Biblical apocalypticism makes us attentive to the divine plan structing the history of salvation, human influence remains an important factor capable of shaping historical processes. Consequently, the model of history operative in Judeo-Christian apocalypses must remain sufficiently open to make space for human agency.

In Isaiah 1:27, the author correlates redemption and destruction with, respectively, righteous penitence and the sinful rebellion of the people. Similarly, in chapter 12 of the Book of Daniel (quoted again in Matthew chapter 24), it is "abomination that causes desolation." But perhaps the most powerful illustration of the importance of human action in an apocalyptic context comes from the Book of Jonah. Read on Yom Kippur, the day of atonement, the subject of the story is *teshuva* – a turning back from a wrong path, which saves the people of Nineveh from destruction.

Contemporary theology, as exemplified by modern Catholicism, is equally attentive to the role played by humanity in shaping the direction of history. As Carmody Grey shows, in the papal encyclical on the environment, "the true arrow of time" is contrasted with its false counterpart, constituted by "the superficial 'technocratic paradigm,' which contradicts creation's deepest nature" (Grey 2019, 9–10). The technocratic paradigm opposes the redemptive arrow of time; consequently, the unravelling of the latter is contingent on a successful struggle with its adversary and the concomitant ecological conversion of both individuals and communities (Francis 2015, §217–219) – reminiscent of the *teshuva* in the Book of Jonah. Similar ideas can be found in Gustavo Gutiérrez's work, where history is structured antagonistically, according to an opposition between forces of liberation and sin; it is precisely this antagonism which makes history undecided, because the shape of the future hinges on overcoming sinful structures and behaviours. The shared argumentative tactic found in the respective theologies of Pope Francis and Gutiérrez consists of splitting history into multiple streams of time. The distinctions between different timelines, in turn, enable theologians to marry together two intuitions: first, the belief in the history of salvation, which is part of God's eternal design, and, second, the sense that history is open and that the earthly future can be decided by human action. Gutiérrez, for instance, specifies three possible meanings of historical processes, corresponding to three types of liberation: political, human, and religious. As he argues, although the streams of history are part of "a single, complex process" (1988, 25), they are nonetheless "deep down different from each other" and "found at different levels" (1988, 103–104). In short, religious liberation – the fulfilment of God's plan of salvation – is both related to and distinct from other historical processes, including political ones, for which humanity finds itself responsible.[1]

We can conclude that Judeo-Christian framing of agency and of history, whether in its older Biblical forms or in its more recent theological expressions, produces insights which resonate with the modern understanding of climate apocalypse: Judeo-Christian texts offer narratives centred on human responsibility, the malleability of historical processes, and the possibility of both catastrophic and emancipatory futures. Moreover, Biblical narratives about the end of the world often involve environmental catastrophes – rendering eco-apocalypse an integral aspect of Judeo-Christian apocalypticism. Revelations chapters 8–9 recount multiple climate disasters: from a fire that consumes a third of the trees on earth, through an extinction of marine life and the poising of waters, to multiple plagues. The Book of Jonah also speaks of natural degradation, although in a more localised way. When Jonah attempts to rest in a shade of plant, God send a worm that makes the plant wither, which exposes Jonah to adverse natural elements so harsh that Jonah wishes he was dead – which God compares to the destruction of the city of Nineveh (Jonah 4:10). The connection between apocalypses and environmental disasters found in Biblical texts can be heard echoed in contemporary Christian theology, explicitly concerned with the modern-day climate crisis. In *Laudato Si'*, for example, Pope Francis writes: "Doomsday predictions can no longer be met with irony or disdain. We may well be leaving to coming generations debris, desolation and filth. . . . The effects of the present imbalance can only be reduced by our decisive action, here and now" (2015, §161).

From revolution to disaster

Judeo-Christian and climate apocalypticisms are further united by a shared semantic connection, which emerges in the late medieval and early modern Europe. My hypothesis is that political movements active at the time, while employing the theoretical tools afforded by Biblical apocalypticism, articulated a recognisably modern meaning of the end of the world, which paired *revolution* and *disaster*. This semantic mutation, in turn, resulted in a (partial) secularisation of apocalypticism and the irreducible overdetermination of the apocalypse as both a hopeful event and a worldly catastrophe – an overdetermination which continues to organise modern eco-apocalyptic discourses. Anders, therefore, was right to point out the significance of revolution within the history of apocalyptic thought; however, in characterising revolt as apocalypse with a kingdom, he missed the fact that political redemption for some was construed as a disaster for others.

The intimate connection between apocalypse and revolution has been noted by Taubes: "if revolution means opposing the totality of this world with a new totality that comprehensively founds anew in the way that it negates . . . then apocalypticism is by nature revolutionary" (2009, 9). Take as an example the Taborites – a radical wing of religious reformers called the Hussites active in 15th-century Bohemia. The Taborites expected the end of the world to take place in February 1420; when this didn't happen, they took it upon themselves to usher the apocalypse by waging a war against the forces of the Anti-Christ loyal to the Catholic Church (Kowalewski 2021). Their armed struggled was coupled with an establishment of the city of Tabor, which overturned the economic, social, and religious relations prevalent at the time. The radicality of the city's arrangement led some contemporary commentators to call the city communist (Kaminsky 1957, 54; Fudge 1998) or even "avowedly anarchic" (Bookchin 1982, 202). In the words of Howard Kaminsky: "[i]f a revolution is defined as the sudden substitution of one social and intellectual world for another, it is clear that Tabor was the first revolutionary society in Europe" (1957, 62).

According to Taubes, the echoes of Taborite political theology can be found in Thomas Müntzer's apocalypticism, whose development coincided with the peasant revolts in 16th-century Germany (2009, 107). Müntzer's theology is particularly significant for the semantic study of apocalypticism – in his polemic with Martin Luther, Müntzer (and Luther) crystalise the tension between utopia and disaster, which continues to inform modern understanding of the apocalypse. In Müntzer's text, beautifully titled "Highly Provoked Vindication and Reply to the Spiritless Easy-Living Flesh in Wittenberg who has Sullied Wretched Christianity with his Falsification and Theft of the Holy Scriptures," the author – in a truly revolutionary fashion – writes:

> Look, our overlords and princes are themselves the breeding ground of extortion, and theft of all kinds; they take possession of all creation, the fish in the water, the birds of the air, all that grows in the soil; everything has to belong to them. Then they spread God's commandment among the poor, saying, God has commended that you shall not steal. But it is of no use to them at all because the princes cause the people, the poor ploughman, the craftsman, all who live in that way to toil and to be ground down to nothing. If he then commits the slightest act of theft, he must hang. To that the lying doctor says, Amen. The lords are responsible for the fact that the poor man becomes their enemy. . . . Therefore, as I say: I must rebel and intend to do so.
>
> *(in Taubes 2009, 112)*

Luther, in a text "Against the Thieving and Murderous Gangs of Peasants," replies by arguing that "rebellion is not a simple case of murder, but a conflagration which ignites a country and devastates it . . . the greatest of all disasters . . . there is nothing more poisonous, harmful, and of the devil than a rebellious person." And so, anybody to is willing to kill a rebellious person, "does well and justly" (in Taubes 2009, 113).

What makes the exchange between Müntzer and Luther fascinating – in addition to enjoyably evocative titles of their pamphlets – is that it marks the emergence of a secular apocalypse we know today from its religious place of origin. Importantly, we catch this birth *in the process*: we can clearly observe how the language connecting redemption, revolution, and disaster becomes articulated and sedimented; how theology becomes secular politics; and how Biblical apocalypses and social revolution pass into one another.

More recent images of revolution continue to evoke its apocalyptic heritage. Mocking "bourgeois consciousness" and its attachment to property, the anarchist Mikhail Bakunin writes in 1873:

> A popular insurrection, by its very nature, is instinctive, chaotic, and destructive, and always entails great personal sacrifice and an enormous loss of public and private property. The masses are always ready to sacrifice themselves; and this is what turns them into a brutal and savage horde . . . they will not hesitate to burn down their own houses and neighbourhoods, and property being no deterrent, since it belongs to their oppressors, they develop a passion for destruction. . . . Revolution requires extensive and widespread destruction, a fecund and renovating destruction, since in this way and only this way are new worlds born.
>
> *(1973, 334)*

Without exaggeration, we can call Bakunin's description of revolution *apocalyptic*; the fact that this adjective imposes itself on us, however, is far from accidental. Bakunin's descriptions repeat the historical overdetermination of apocalypse, and its semantic bifurcation, and its bifurcation forged in revolutionary conflicts and informed by theology – where the image of redemption becomes inseparable from political violence and where the birth of the new world requires a catastrophic ruin of the old one. This political–theological heritage continues to inform today's attitudes – think here of how violent riots, burning cars, and looted shops can appear to some as a "conflagration which ignites a country" or as an image of nearly divine retribution for years of oppression.

As I will show in the next section, contemporary environmental discourse takes up and transforms the historical overdetermination of the apocalypse by approaching it through a prism of war. The latter concept becomes split into a disastrous "war against the earth" and a revolutionary "war for the earth." This conceptual bifurcation, however, wouldn't be possible without the sedimented material left behind by the political apocalypticism of the past (and its critics); Western environmentalism, therefore, is indebted to the theo-political debates accompanying the birth of modernity – often without knowing it.

War and the earth

In *Savage Ecology*, Jairus Victor Grove links "industrialized war, capitalism, and ecological destruction," arguing that the "elite-driven Euro-American geopolitics of industrialized war and capitalism made ecocide that is now a global historical fact" (Grove 2019, 10–11).

> The two most important components of industrialization – interchangeable parts and the assembly line – were developed because of the demands of larger and larger armies, not larger and larger civilian markets. Furthermore, the demand for American industrialization was not the result of an "invisible hand" but a directive of the War Department to create operations for arms production using interchangeable parts.
>
> *(Grove 2019, 106)*

Both industrialisation and war are responsible for the direct devastation of the environment. The extraction of raw materials and burning of fossil fuels, which animates industrialism, destroys natural spaces while the process of production pollutes air and water. War – made possible by the industrial production of weapons – also transforms landscapes. Raids, bombing, and scorched earth tactics extend the effects of industrialisation. Russian invasion of Ukraine is a case in point: not only is it heavily financed by Russian exports of natural gas and other fossil fuels, it also results in an extensive environmental damage in Ukraine. "In the most recent spring season (2023), a considerable portion of Ukrainian land was uncultivated, due to contamination by explosive objects, with approximately 30% of the country's total territory affected" (Zagoruichyk et al. 2023).[2]

The famous painting "Coalbrookdale by Night" by Philippe Jacques de Loutherbourg pictures the eco-apocalyptic connection between industrialisation and war, where the fires of the steel plant, set against the background of a dark night, can be easily confused with an aftermath of a battle. However, it is the effects of nuclear conflict which illustrate most explicitly the apocalypticism of war:

> Even the smallest of nuclear weapons . . . exploding in modern megacities would produce firestorms that would build for hours, consuming buildings, vegetation, roads, fuel depots, and other infrastructure, releasing energy many times that of the weapon's yield.
>
> *(Mills et al. 2014, 161)*

It is, therefore, unsurprising that contemporary Western environmentalism, sensitive to our apocalyptic condition, has taken up the language of war. As Malm observes: "[m]etaphors of war roll off the tongue *when life and death on a mass scale are at stake* and *the situation demands extraordinary mobilisation to survive*" (2020, 154). The concept of war has circulated for decades in ecological discussions. To name just a few examples – in an article published in *the Guardian*, Joseph Stiglitz states: "[t]he climate crisis is our third world war" (Stiglitz 2019); a few years earlier, Vandana Shiva delivers a speech published under the title "Time to end war against the earth" (Shiva 2010); while in 1996 Polish anarcho-punk band Włochaty names its album *Wojna przeciwko Ziemi (War against the earth)*. Here, again, nuclear war becomes a paradigmatic point of reference; as Rens van Munster puts it:

> Not only do nuclear weapons continue to threaten the possibility of a (human) future, but also the current climate crisis is increasingly understood through the prism of a sixth global extinction event. We continue to live in the final age.
>
> *(2023, 17)*

The notion of "war against the earth" can be contrasted with the sentiment of "war for the earth" expressed by the more radical wing of Western environmentalists, which draws on the semantic link between war and revolution found in Marxism (echoing medieval and early modern apocalypticism discussed in the previous section). Andreas Malm proposes to learn from the experience of the Bolsheviks, albeit in a qualified manner, and to revive the notion of *war communism* – or ecological Leninism – whose "basic make-up must harbour predisposition for emergency action and openness to some degree of hard power from the state" (Malm 2020, 153). Similarly, in *Climate Change as Class War: Building Socialism on a Warming Planet*, Mathew T. Hubner advocates for class struggle over the access to the ecological needs of life in the sphere of ownership and control of production (2022). Clearly, the apocalyptic theologies of war and revolution developed by Müntzer and the Taborites continue to exercise an influence on radical environmentalism – albeit indirectly, via Marx and Marxism.[3]

Silencing the apocalypse and climate denialism

The connection between (eco-)apocalypse, revolution, and war, explored in the preceding sections, attests to a process of semantic expansion, that is of "swallowing up" of meanings by apocalyptic concepts. Importantly, a closer look at war and revolution can also illustrate how the semantic content of apocalyptic notions can *shrink* and how the overdetermined meanings within the orbit of apocalypticism can become *silenced*. As I suggest, this process of diminishing or obscuring apocalyptic meanings underlies the phenomenon of climate denialism.

René Girard identifies two processes by which apocalyptic meanings can be covered over. The first one consists of increasing violence to a point at which it drowns out the previous apocalypse. For example, the introduction of totalitarianism in the 20th century was "a way of *not wanting to see* what happened at Verdun, of wishing the apocalypse away by speeding up its course" (2010, 40). The second process consists of abandoning the very idea of apocalypse the moment the notion becomes realised:

> It was probably beginning with Hiroshima that the idea of the apocalypse completely disappeared from the Christian mind: Western Christians, French Catholics in particular, stopped talking about the apocalypse just when the abstract became real, when reality began to match the concept.
> *(Girard 2010, 64)*

A particular modification of the latter process has been identified by Andrzej Leder, who notes how sometimes correct concepts are unable to materialise, despite the fact that the apocalyptic reality calls for it. Leder's example is one of recent Polish history. The period of 1939–1956 in Poland, that is Nazi occupation and Stalinism, marks a type of social revolution: "cruel, brutal, imposed form the outside, but revolution nonetheless" (2014, 7 *my translation*), which violently altered the fabric of Polish society. However, the *concept* of this revolution is strangely absent from Polish national consciousness. This is because, Leder argues, the revolution was taken away from the Polish political subject by German and Russian powers; since it was carried out by others, the violent upheaval of 1939–1956 was experienced as a "nightmare" happening *to* the subject and so without the identification of the latter with decisions and actions responsible for the changes. In short, the revolution was missed or stolen. Consequently, it remains conceptually unacknowledged, despite the fact that, as Leder suggests, contemporary Poland is a direct effect of this brutal social revolution.

The fact that apocalypse can be silenced, that is it can be drowned out or it can lack the concepts to expresses it, can help us understand the phenomenon of climate denialism in its various guises. For some, the environmental end of the world may be a nightmare happening *to* them, and so it will remain excluded from their consciousness and therefore inexpressible through a concept of climate apocalypse. Here, the apocalypse would be stolen or missed – the extreme weather events would appear as occurring as if all by itself. For others, the reality of climate apocalypse can be veiled by a constant increase of apocalyptic events, drowning out catastrophes with more catastrophes, ultimately leading to a disappearance of adequate concepts through which to grasp one's surroundings. It could be hypothesised that this process is exploited by the oil industry, whose strategy is to simultaneously increase apocalyptic events through the exploitation of fossil fuels and to argue against the concepts designating the environmental end of the world.

Crisis, catastrophe, extinction

Although other notions have a significant presence in the contemporary ecological discourse, Western environmentalism is dominated by a triad of concepts, which mark the three main inflections of climate apocalypse: the latter can be referred to as a *crisis*, a *catastrophe*, or an *extinction event*. As I will show in this section, the difference between these concepts is one of degree, not of kind; as a result, the relationship between these three terms in the context of eco-apocalypse attests to the constitutive blurring of semantic boundaries proper to apocalypticism. Furthermore, the analysis of extinction will lead us back to Biblical apocalypses, illustrating, once more, the continual relevance of the theological heritage for the understating of climate apocalypticism.

As Reinhart Koselleck argues, the concept of crisis "indicates that point in time in which a decision is due but has not yet been rendered . . . the concept is applied to life-deciding alternatives meant to answer question about . . . what contributes to salvation or damnation" (2006, 361). However, the centrality of decision doesn't protect the concept of crisis from being combined with other terms; crisis "is often used interchangeably with 'unrest,' 'conflict,' 'revolution,' and to describe vaguely disturbing moods or situations" (Koselleck 2006, 399).

A similar plurivocity can be found on the side of concepts which are used to describe the effects of climate change often in counter-distinction to the notion of crisis. Recently, Jonathon Catlin proposed to conceive global warming as "slow catastrophe" situated in between the optimism characteristic of crisis-rhetorics and the apocalyptic pessimism proper to prophecies of extinction. However, when engaging with Catlin's argument, it soon becomes apparent that the distinctiveness of slow catastrophe is grounded not in its ability to separate itself from crisis or extinction but rather in its power to combine the elements of crisis and extinction, as well as other conceptual oppositions, *within itself*. Slow catastrophe is a "a dialectical middle ground" which "holds both the danger and opportunity of climate emergency in view, conceptualizing it as both sudden *and* slow, continuous *and* discontinuous, and structural *and* evental" (Catlin 2023, 64). Consequently, the distinguishable feature of the notion of crisis – namely, decision – becomes subsumed into more expansive concept, which designates also the object of the decision: the environmental catastrophe.

Can a similar blurring of semantic boundaries be found in the case of extinction? Arguably, the effects of extinction would enable it to escape the subsumption under other, less extreme concepts. A mass extinction event, for example, would be at odds with the notion of a crisis. While the latter necessitates a decision over a future, and implies a set of values which guide this decision, extinction would signify the anticipation of absolute death and with it the disappearance of both the object of the decision and the normative conditions of decision-making. As Ray Brassier argues, the possibility of a future extinction event can reflect back on the current moment – annihilation is a trauma (albeit an expectational one, insofar as it is located in the future) which has the ability to undermine life and the values which orient it *today*. Because extinction will have happened, "[e]verything is dead already" (Brassier 2007, 223). On Brassier's nihilistic reading, extinction would become a catastrophe incapable of functioning as a "dialectical middle ground," since it would refuse any commonality with crisis-optimism and the decision over the future a crisis involves.

However, it is possible to invert Brassier's nihilism – which is exactly what Biblical apocalypticism proposes. In the words of Stefan Skrimshire, in Judeo-Christian apocalypses

> value is to be found in present existence precisely because its continuation in human terms not only can never be guaranteed, but moreover it has its "end" always in sight. The apocalyptic imagination can be seen as an attempt to provide depth to one's commitment in the present *in light of*, rather than *in spite of*, its transitory and finite nature.
>
> *(2023, 177)*

On the reading based in Biblical apocalypses, extinction would still constitute a future trauma reflecting back on the normative conditions of the present; however, instead of resigning ourselves to nihilism, we can view the impact of extinction on our normative categories as a deconstruction – a call for a renewed ethics capable of facing up to the potential death of humanity. "Apocalyptic ethics states: prepare yourself *as if* the end was coming now, as if the 'end times' were now" (2023, 181). Understood this way, the horizon of extinction enacts "the revelation of common mortality" (2023, 181), a disclosure of our existential condition which forces us to respond to our finitude. The revelation of extinction, therefore, consists of uncovering that apocalypse is an inescapable point of reference for our ethics, epistemology, and ontology.[4] The way we act and think in light of who we are can't be separated from apocalypticism. Importantly for our purposes, understanding extinction as revelation draws it back into the semantic relationship with both crisis and catastrophe. Any existential decisions over the dangers and opportunities of a given disastrous situation are situated against the backdrop constituted by the possibility of the absolute death of life. Here, extinction becomes a catastrophic horizon of a crisis situation, which fuels, instead of suppressing, decision, thought, and action.

Revelation: mystery, matter, and ideology

At the end of the previous section, our analysis of the orbit of (eco-)apocalypticism has arrived at the notion of *revelation* – the etymological sense of apocalypse. Revelation has played a double function in the context of our discussion; it named the power of the end of the world to uncover the ethical, epistemic, and ontological structures of our being, and it constituted a bridge between religious and secular apocalypticism (both the Book of Revelation and climate extinction tell us something about the finite conditions of human thought and action). The third effect of revelation, which is the subject of this section, is its ability to expose the role of ideology in the constitution of the semantic field of apocalypticism.

Here, I will contrast two philosophical approaches to the concept of revelation – one *mystical* and the other *materialist*; I will then show how neither the mystical nor the materialist is able to uncover that which remains hidden in any apocalypse, including in its environmental instantiation – namely, the ideological condition of thinking and acting apocalyptically, expressed as a demand to ensure the reproduction of our conditions of existence.

The first, mystical approach, developed by Jean Vioulac, takes as its starting point the prophet Daniel – a messenger of God who, famously, struggled to understand his own visions. While Althusser calls Daniel an "idiot" (1993, 216), Vioulac believes that Daniel's lack of understanding provides a clue enabling us to identify the essence of revelation. What Daniel stumbled upon, Vioulac argues drawing on Heidegger, is the ground of phenomena, which *withdraws* as it allows entities to appear. Daniel encounters "the presence proper to absence . . . *the phenomenality proper to disappearance*" (Vioulac 2021, 38), which, via Heidegger, Vioulac equates with the retreating conditions of phenomenality. In other words, Daniel has every right to be confused because the very ground of meaning on which his vision is built seems to be an absence – that which makes things appear, dissimulates itself. Revelation, therefore, in generating images and meanings, enables our contact with the withdrawing ground, the *"glaring absence,"* experienced in and as mystery (2021, 38). Additionally, the confusion experienced by Daniel is caused by the fact that the absence of the ground given in revelation has a catastrophic effect for the categories of presence, truth, and the world: presence is no longer primary, since it presupposes absence as its condition; truth can no longer rely on presence of things, and thus it is undermined; finally, the world structured around truths and present objects collapses (Vioulac 2021, 55).

The second philosophical approach to revelation, which can be seen as offering a *materialist* counterpoint to Vioulac's account, can be found in the work of Evan Calder Williams. Similarly to his Heideggerian counterpart, Williams views apocalypse in terms of its revelatory power: "apocalypse is not the clarification itself but a wound of the present that exposes the unseen"

(2011, 6). However, in contrast to Vioulac, Williams equates the unseen with the material conditions of the present social, political, and economic order. Apocalypse, therefore, enables us to grasp:

> how the global economic order and its social relations depend upon the production and exploitation of the undifferentiated . . . hellish zones of the world, whole populations destroyed in famine and sickness, "humanitarian" military interventions, the basic and unincorporable fact of class antagonism, closure of access to common resources, the rendering of mass culture more and more banal, shifting climate patterns and the "natural" disasters they bring about, the abandonment of working populations and those who cannot work in favor of policies determined only to starkly widen wealth gaps.
>
> *(Williams 2011, 8)*

While I am sympathetic to Williams's reading of apocalyptic revelation, I believe that it misses something important – and it does so, I believe, because it is *insufficiently* Heideggerian. A materialist concern with *beings* (famine, military interventions, disasters etc.) runs the risk of remaining blind to that which dissimulates itself in making entities appear. Importantly, however, we don't have to follow Vioulac and label this absent ground of phenomena a *mystery*; rather, in line with my argument in the previous chapter, we can attribute this specific type of non-presence to *ideology*. The ideological concern with reproduction can't be reduced to the materialist revelation of entities, because – as we have seen in Chapter 1 – it is the condition which orients our experience of the worldly phenomena as apocalyptic; equally, ideology can't be described only as an absent ground of phenomenality, since it introduces a positive content, a demand to reproduce our conditions of existence. Thus, if apocalypses – and the terms which circulate in their orbit – reveal some things so far hidden, then what they uncover is the force of ideology, which withdraws as it produces experiences and notions.

Ideological fixing

We can begin to construct a theoretical model capturing the functioning of the semantic orbit of apocalypticism with the help of Michael Freeden's conceptual approach to the study of ideology and Slavoj Žižek's account of ideological struggle.

As Freeden notes, ideologies possess a "morphology" of *core*, *adjacent*, and *peripheral* concepts. We should be careful, however, not to interpret the core of ideology as "single constituent concept," nor as "a structurally fixed and substantively permanent set of concepts." Rather, the core is "a flexible and empirically ascertainable collection of ideas, fashioned by social conventions" (2006, 84). Ideologies are semantic fields, "open-bordered and inconstant mutation" (2006, 82), in which elements interact and transform each other.

Importantly, particular ideologies are engaged in an ideological struggle "over the socially legitimated meanings of political concepts and the sustaining arrangements they form, in an attempt to establish a 'correct' usage" (Freeden 2006, 77). It is the ideological struggle which can help us understand both the specific morphology of the ideological field and the changes it has undergone over time. As Žižek explains, ideological struggle aims to "quilt" floating notions:

> The "quilting" performs the totalization by means of which this free floating of ideological elements is halted, fixed – that is to say, by means of which they become parts of the structured network of meaning. If we "quilt" the floating signifiers through "Communism," for example, "class struggle" confers a precise and fixed signification to all other elements. . . . What is at stake in the ideological struggle is which of the "nodal points," *points de capiton*, will totalize, include in its series of equivalences, these free-floating elements.
>
> *(1989, 95–96)*

Note that in this passage, Žižek identifies *two* types of nodal points – *communism* which gathers together the signifiers and *class struggle* which endows signifiers with meaning. I will come back to this twofold process of quilting later. For now, it is sufficient to say that nodal points are privileged concepts which unify and bestow a specific arrangement on a given ideological field. The goal of ideological conflict, therefore, is to "fix" the signifier which will perform the function of the *point de capiton*: "the word to which 'things' themselves refer to recognize themselves in their unity" (Žižek 1989, 105).

Interestingly, the process of fixing *apocalyptic* signifiers by a nodal point has been intuited by Vioulac. The following passage captures both the complex conceptual morphology of apocalypticism and the power to quilt floating signifiers by the notion of the apocalypse:

> In this way, our epoch is the catastrophe of ontological truth. This catastrophe is not unilateral, however; as the collapse of truth, it dislocates the conditions of manifestation, which allows, for the first time, what these conditions expressly dissimulate to be glimpsed. . . . This revelation abruptly convicts truth itself of error and shows an errancy in its destiny. The revelation of the mystery is a crisis. . . . Such an event must be circumscribed by the concept of apocalypse.
>
> *(Vioulac 2021, 52)*

Returning to Freeden and Žižek, we should note that they diverge in their understanding of the content of the conceptual nodal point. Freeden argues for an *ineliminable feature* of a notion, whose "absence would deprive the concept of intelligibility and communicability" (2006, 62). Žižek, by contrast, argues that the ineliminable feature of a core notion is an *absence* – it can signify only by *reflecting back* the words, contexts, and things it unifies. In other words, its presence is functional not semantic: the nodal point quilts the ideological field, and in so doing it mirrors back the meanings of other signifiers. Consequently, the core concept is "in its bodily presence nothing but an embodiment of a certain lack" (Žižek 1989, 110).

If Žižek was correct, his claim would have direct consequences for how we conceive of the structure of the semantic field of ideology – including the conceptual structure of (eco-)apocalypticism. Signifiers, things, and contextual circumstances would derive their meaning from the *point de capiton*; the core concept, however, lacking a meaning of its own, would refer us back to signifiers, things, and contexts. Thus, in place of an ineliminable feature of a concept, we find an *ineliminable absence*, which hides itself in the interminable back-and-forth reference between the core, peripheral, and adjacent notions, as well as their contexts and objects. If we apply Žižek's model to apocalyptic signifiers quilted by climate apocalypse, we can suggest that any positive definition of eco-apocalypse could be constructed only by borrowing content from the related notions it gathers together. Conversely, when in another historical context apocalypticism is fixed by religious apocalypse, the Biblical end of the world would mirror the meanings of adjacent concepts which surround it – including, as we have seen, climate apocalypse.

Although the semantic analysis presented in this chapter supports this conclusion, I nonetheless believe that Žižek's theory must be qualified by noting the positive content of ideology, touched on in the previous section. As I argued in Chapter 1, apocalyptic notions possess heterogeneous *referents* and a shared *sense* – the latter expressing the ideological demand to identify threats to the reproduction of our conditions of existence. My contention is that it is the ideological sense which provides the, for the most part, *unconscious* nodal point which fixes floating signifiers. By virtue of hiding itself within the referential relationship between signifiers and things, the ideological guarantee of meaning can, as Žižek suggests, appear as a lack. However, this doesn't prevent it from injecting apocalypticism with *positive* content: a specific demand for reproduction which constitutes the ineliminable (and often unconscious) feature of apocalyptic notions.

My hypothesis is that, in fact, we can identify two processes of quilting: the first achieved by the dominant concept in the ideological field, for example climate apocalypse, which reflects and organises other signifiers; and the second carried out by ideological sense. In fact, Žižek himself intuited this distinction when he split the fixing function between *communism* and the *class struggle*. However, his attachment to the idea of a lack made him miss the positivity of the ideological demand, which injects the semantic field of ideology with plenitude of meaning, even if this meaning is for the most part hidden. Thus, the work of the conceptual nodal point – for example eco-apocalypse – which mirrors adjacent meanings and is responsible for the articulation of a particular morphology of the semantic field (distributing the roles of core and peripheral notions) takes place in parallel to the quilting enacted by the ideological demand. The latter introduces a unifying sense, a shared – though often implicit – concern, which enables the communication between the concepts constitutive of apocalypticism. Consequently, to answer the question of what climate apocalypse means, we must examine the concepts circulating in its orbit – which eco-apocalypse both quilts and reflects – and the ideological demand for reproduction which unifies the semantic field of apocalypticism. The interaction between these two quilting functions generates a labyrinth-like structure in which overdetermined notions exchange meanings and positions as a result of ideological struggle.

Notes

1 What makes theologies of Gutiérrez and Francis capable of speaking to an apocalyptic problematic is a threefold relationship to historical apocalypticism. First, their concern with the fulfilment of history and human actions can be read as a direct engagement with themes found in Biblical apocalypses; second, presenting history as structured by struggle with an adversary is equivalent to taking up and rephrasing the figure of the Anti-Christ, whose defeat in apocalyptic narratives is necessary for the arrival of New Jerusalem; finally, the question of earthly liberation feeds not only on the emancipatory sentiments of the 20th century but also on the revolutionary fervour of apocalyptic movements found in previous centuries.
2 For an analysis of the connection between "the destruction of Palestine and the destruction of the Earth," see Malm (2024).
3 For a discussion of the importance of apocalypticism for Marx, see Taubes (2009).
4 This intuition is echoed by Brassier, who notes how extinction "turns thinking inside out, objectifying it as a perishable thing in the world like any other" (2007, 229).

Bibliography

Althusser, L. (1993) *The Future Lasts a Long Time*. Trans. R. Veasey. London: Chatto & Windus
Anders, G. (2019) "Apocalypse Without a Kingdom," *E-Flux Journal 97*. Available online: www.e-flux.com/journal/97/251199/apocalypse-without-kingdom/ [Accessed 20/10/2024]
Bakunin, M. (1973) "Statism and Anarchy," in *Bakunin on Anarchy*, S. Dolgoff (Trans. & Ed.). London: George Allen & Unwin Ltd
Bookchin, M. (1982) *The Ecology of Freedom: The Emergence and Dissolution of Hierarchy*. Palo Alto: Cheshire Books
Brassier, R. (2007) *Nihil Unbound: Enlightenment and Extinction*. London: Palgrave Macmillan

Catlin, J. (2023) "Slow Catastrophe: A Concept for the Anthropocene," in *The Environmental Apocalypse: Interdisciplinary Reflections on the Climate Crisis*, J. Kowalewski (Ed.). Abingdon: Routledge

Francis, P. (2015) *Laudato Si': On Care for Our Common Home*. London: Catholic Truth Society

Freeden, M. (2006) *Ideologies and Political Theory: A Conceptual Approach*. Oxford: Oxford University Press

Fudge, T.A. (1998) "'Neither Mine Nor Thine': Communist Experiments in Hussite Bohemia," *Canadian Journal of History* 33(1), pp. 25–47

Girard, R. (2010) *Battling to the End: Conversations with Benoît Chantre*. Trans. M. Baker. East Lansing: Michigan State University

Grey, C.T.S. (2019) "Time and Measures of Success: Interpreting and Implementing *Laudato Si*," *New Blackfriars* 101(1091), pp. 5–28

Grove, J.V. (2019) *Savage Ecology: War and Geopolitics at the End of the World*. Durham, NC: Duke University Press

Gutiérrez, G. (1988) *A Theology of Liberation: History, Politics, and Salvation*. Trans. Sr. C. Inda & J. Eagleson. Bungay: SCM Press Ltd

The Holy Bible. (2011) *New International Version*. Palmer Lake: Biblica

Hubner, M.T. (2022) *Climate Change as Class War: Building Socialism on a Warming Planet*. London: Verson

Kaminsky, H. (1957) "Chiliasm and the Hussite Revolution," *Church History* 26(1), pp. 43–71

Koselleck, R. (2006) "Crisis," *Journal of the History of Ideas* 67(2), pp. 357–380

Kowalewski, J. (2021) "The Transfigurations of Spacetime: The Concept of Tabor in the Hussite Revolution and Its Implications for Philosophy of History," *Praktyka Teoretyczna* 1(39), pp. 161–186

Leder, A. (2014) *Prześniona rewolucja. Ćwiczenie z logiki historycznej*. Warszawa: Wydawnictwo Krytyki Politcznej

Malm, A. (2020) *Corona, Climate, Chronic Emergency: War Communism in the Twenty-First Century*. London: Verso

Malm, A. (2024) "The Destruction of Palestine Is the Destruction of the Earth," *Verso*. Available online: www.versobooks.com/blogs/news/the-destruction-of-palestine-is-the-destruction-of-the-earth?srsltid=AfmBOopuQOOk9ShFIyV8xuBgJDu0Av4w-YWz4saS2NM4XlZMMwb5Aj8h [Accessed 4/12/2024]

Mills, J.M., Owen, T.B., Lee-Taylor, J., Robock, A. (2014) "Multidecadal Global Cooling and Unprecedented Ozone Loss Following a Regional Nuclear Conflict," *Earth's Future* 2(4), pp. 161–176

Scholem, G. (1971) *The Messianic Idea in Judaism: And Other Essays on Jewish Spirituality*. Trans. M.A. Meyer et al. New York: Schocken Books

Shiva, V. (2010) "Time to End War against the Earth," *The Age*. Available online: www.theage.com.au/opinion/society-and-culture/time-to-end-war-against-the-earth-20101103-17dxt.html [Accessed 20/10/2024]

Skrimshire, S. (2023) "Apocalyptic Time and the Ethics of Human Extinction," in *The Environmental Apocalypse: Interdisciplinary Reflections on the Climate Crisis*, J. Kowalewski (Ed.). Abingdon: Routledge

Stiglitz, J. (2019) "The Climate Crisis Is Our Third World War. It Needs a Bold Response," *The Guardian*. Available online: www.theguardian.com/commentisfree/2019/jun/04/climate-change-world-war-iii-green-new-deal [Accessed 20/10/2024]

Taubes, J. (2009) *Occidental Eschatology*. Trans. D. Ratmoko. Stanford: Stanford University Press

Van Munster, R. (2023) "Nuclear Weapons, Existentialism, and International Relations: Anders, Ballard, and the Human Condition in the Age of Extinction," *Review of International Studies* 49(5), pp. 813–831

Vioulac, J. (2021) *Apocalypse of Truth: Heideggerian Meditations*. Trans. M.J. Peterson. Chicago: The University of Chicago Press

Williams, E.C. (2011) *Combined and Uneven Apocalypse: Luciferian Marxism*. Winchester: Zero Books

Zagoruichyk, A., Savytskyi, O., Kopytsia, I., O'Callaghan, B. (2023) *The Green Phoenix Framework: Climate-Positive Plan for Economic Recovery in Ukraine*. Oxford Smith School of Enterprise and the Environment – Working Paper No. 23–03. Available online: www.smithschool.ox.ac.uk/sites/default/files/2023-06/The-Green-Phoenix-Framework-a-climate-positive-plan-for-economic-recovery-in-Ukraine.pdf [Accessed 21/10/2022]

Žižek, S. (1989) *The Sublime Object of Ideology*. London: Verso

3
WHY BEING IS ON NOBODY'S SIDE

The politics and ontology of climate apocalypse

In the 1960s, Posadas and his Trotskyist followers saw the nuclear war as the means to destroy capitalism. Consequently, they began calling for nuclear apocalypse as a necessary step towards the lasting victory of socialism (Gittlitz 2020). Today, apocalypticism seems to be the domain of right-wing conspiracists; as Neal Curtis notes, "to 'be redpilled' is to share in a communal uncovering of dangerous – if not actually evil – forces that threaten the annihilation of worlds" (2022, 98). These two examples – one from the far left and the other from the far right of the political spectrum – may give the impression that apocalypticism is a phenomenon found on the extreme edges of politics. This conclusion, however, wouldn't be accurate. Apocalyptic visions of the end of history, The Invisible Committee tells us, are integral to the ideologies of "[s]ocialists, liberals, Saint-Simonians, and Cold War Russians and Americans," insofar as all express "the same neurasthenic yearning for the establishment of an era of peace and sterile abundance . . . an earthly paradise organised on the model of a psychiatric hospital or a sanatorium" (2015, 38–39). On this reading, most contemporary political ideologies would be (crypto-)apocalyptic, at least to some extent. The Invisible Committee's hypothesis seems confirmed by the fact that Posadas nuclear prophecies merely intensify the anticipation of an economic apocalypse proper to mainstream Marxism and that today's right-wing conspiracy theories "have started to dovetail with mainstream conservatism" (Curtis 2022, 97).

Although apocalypticism can be found expressed equally by both fringe and more conventional ideologies, it nonetheless remains possible to draw political distinctions between different types of apocalypse. According to The Invisible Committee, apocalyptic prophecy is often "pronounced only in order to summon the means of averting it, which is to say, most often, the

DOI: 10.4324/9781003348511-4

necessity of government" (2015, 35–36). This is not the whole of the story, however; as we have seen in the preceding chapter, apocalypticism also feeds projects where the figure of the end of the world is tied to revolutionary goals, which directly oppose dictatorial politics. To capture the distinction between authoritarian and anti-authoritarian apocalypticisms, Jacob Taubes has famously proposed to differentiate between apocalypse *from above*, which designates the employment of apocalypticism by powers that be for the preservation of the status quo, and apocalypse *from below*, which informs emancipatory and revolutionary projects aimed at radical transformation of the world (2013). In the Althusserian vocabulary, apocalypses from above and from below would constitute, respectively, dominant and the subordinate ideological tendencies, engaged in an ideological struggle (Althusser 2011).

The aim of this chapter is to test the application of the category of apocalypse from below to contemporary climate politics. At first glance, apocalypse from below would be a useful designation for popular, bottom-up, and democratic ecological ideologies with apocalyptic sensibilities. However, a closer look reveals that the Taubesian category is unable to capture the complexity of ideological positions expressed by Western environmentalism. Despite appearances, the label of apocalypse from below can't be accurately applied to *most* ecological ideologies, however democratic or revolutionary they seem to be. Furthermore, as I argue, ideologies which do instantiate Taubesian apocalypse from below operate with an ontological commitment to a plastic and changing being, tending towards the end of the world; as a result, apocalypse from below inadvertently perpetuates an ontology which fuels and justifies the authoritarianism of apocalypse from above.

I conclude this chapter by suggesting that to save the spirit of apocalypse from below, we must abandon the ontology which underlies both sides of the Taubesian distinction. Paradoxically, asserting that the world *may* or *may not* end enables us to refocus climate apocalypticism by making it more sensitive to multiple and localised apocalypses and to the stakes of environmental politics, which can no longer find solace in the ontology of necessary change, nor in the straightforward distinction of apocalypse from below and apocalypse from above.

Apocalypse from above, apocalypse from below

Let's begin with Carl Schmitt. The German jurist was famously interested in the figure of the katechon – the restrainer found in Paul's second letter to the Thessalonians. In Schmitt's political theology, the katechon is "the historical power to *restrain* the appearance of the Antichrist" and hold back the end of the world (2006, 59–60). While Schmitt identifies the katechon with imperial power, Stefan Skrimshire finds this figure as implicitly present in Schmitt's earlier writings, in the form of "the principle of dictatorship"

(2019, 531). Katechonic politics "means living in end-times in the sense of a patient endurance of the status quo and the resistance of its collapse to the very last minute, with forms of authoritarian political measures" (Skrimshire 2019, 532).

Taubes famously positions his political theology in opposition to Schmitt's. "Carl Schmitt thinks apocalyptically, but from above, from the powers that be; I think from the bottom up" (Taubes 2013, 113). And so, if Schmitt is the prophet of counter-revolution, Taubes's apocalypse from below is explicitly revolutionary (Taubes 2013, 11). On this reading, Skrimshire observes, "[e]schatological conflict is not to be understood as the perpetual delay of chaos by means of empire, but rather a vision of the end of empire itself" (Skrimshire 2019, 532).

Skrimshire has suggested that the distinction between apocalypses from above and from below can help us map the different forms of climate politics. In addition, the legacy of apocalypticism expressed by Taubes can also inform eco-activism, which opposes the "climate leviathan" and its top-down environmental policies: "what is needed is a recovery of those theological traditions for which living in the end-times calls for a radically new vision of politics itself" (2019, 533).

> The need for a *katechon* figure, and the principle of dictatorship, may well become increasingly sought in an age of climate emergencies. This is a new, and troubling development. . . . In light of the catastrophic failures of political efforts to safeguard the future of human societies, a theology over-emphasising original sin, and the need for an external force to step in to save us from ourselves, even if it means suspending liberties, seems increasingly plausible to some. . . . Movements like Extinction Rebellion have been aware of this danger, and made explicit their aim to renew democracy in the form of "citizens assemblies" in the enacting of any state of climate emergency. There is, moreover, greater distrust than ever before that political leaders will spearhead the global changes needed to meet the crisis.
>
> *(Skrimshire 2019, 531)*

However, understanding eco-apocalyptic politics as split according to the categories of apocalypse *from above* and apocalypse *from below* can be problematised. As Skrimshire himself observes, the figure of the katechon is ambiguous (2019, 530). The ambiguity stems from the fact that the concept of the restrainer has two distinct aspects: the authoritarian principle of dictatorship and the power capable of withholding the end of the world. While for Schmitt these two elements appear together, in climate politics they are often separated. For example, eco-activists may advocate *against* the principle of dictatorship; however, insofar as their goal is to avert the disaster related to biodiversity loss and climate change, they assume the role of the restrainer. Consequently, eco-activism couldn't be placed comfortably in the category of apocalypse from below. While activists' critique of governments can be classified as emancipatory, insofar as part of their apocalyptic logic is preventative, they draw from katechonic political theology – which places them partially in the category of apocalypse from above.

Andreas Malm is a good example here. In *How to Blow Up a Pipeline*, Malm advocates for grassroots emergency tactics, including "the damage and destruction of property" (2020b, 109). The goal is "to consciously intervene so as to stop this civilization from destroying itself by destroying the foundation on which any organised life must stand" (2020a, 118). But, to effectively withhold the climate apocalypse, the popular campaigns targeting property must be supplemented by additional measures, for example mandatory global veganism, "audit of supply chains and import flows," paying "for tropical areas previously devoted to norther consumption to be reforested and rewilded," and "ban on importing mean from countries in or bordering on the tropics." He concludes that the withholding of eco-apocalypse "would require some coercive authority" (2020a, 151). "Clearly it would be the state that would have to do this. No mutual aid group in Bristol could even hypothetically initiate a programme of this kind" (2020a, 131). Malm's dual power theory, combining popular campaign of property destruction and top-down "ecological Leninism," clearly merges elements of both apocalypse from below and apocalypse from above. Although other eco-activists may reject the need for eco-Leninism, insofar as their goal is to influence government policies, they are committed to some form of top-down imposition to restrain the forces of climate apocalypse – and, therefore, they align themselves with apocalypse from above, despite their avowed popular-democratic ethos.

Furthermore, it is unclear if the apocalypticism from below as expressed by Taubes can at all inform eco-activism. This is because Taubes refuses both aspects of katechonic political theology: he is against the authoritarian principle of dictatorship and the preventative work of the withholding power. Taubes wants to unleash the chaos of apocalypse; as he famously puts it, "I can imagine as an apocalyptic: let it go down. I have no spiritual investment in the world as it is" (2004, 103). Although the echoes of his apocalypticism reverberate among eco-activists who take seriously the notion of societal collapse, even they display an *investment* in the world.

Jem Bendell, a theoretician associated with Extinction Rebellion, sees collapse of societies as a necessary effect of climate change. Faced with the prospect of eco-apocalypse, humanity must adapt to the post-apocalyptic world (2020). Bendell's deep adaptation offers the means to ensure the survival of humanity prepared for the collapse. Here, apocalypticism is only secondary to the investment in the future of both nature and humanity and the preservation (or even improvement) of our conditions of existence. We are letting the world go down only in part and only because of our more fundamental investment in its future existence.

It may appear that the only subjects capable of truly embodying Taubes's sentiment are climate denialists, insofar as their lack of acknowledgement of the climate apocalypse is actually letting the world burn. However, as Bruno Latour observes, even climate denialists express a deep attachment to the world. The Western climate sceptic believes that history has already ended, and we have reached the ultimate epoch; consequently, the idea that their New Jerusalem is under threat seems to fly in the face of their investment in the stability of the eternal reproduction of their conditions of existence.

> Telling Westerners – or those who have recently become Westernized, more or less violently – that the time has come, that their world has ended, that they have to change their way of life, can only produce a feeling of total incomprehension, because, for them, the Apocalypse *has already taken place*. They have already gone over to the other side. The world of the beyond has been achieved – in any case for those who have become wealthy. They have already crossed the threshold that puts an end to historicity. They know, they hear, but deep down *they do not believe it*. Here is where we have to seek the fundamental source of climate skepticism, I believe.
>
> *(Latour 2017, 206)*

Taubesian apocalypse from below, therefore, would seem irreconcilable with environmentalism in its diverse guises, since even the climate denialist confesses to the investment in the world.

Nonetheless, I believe that it is possible to use Taubes's category of apocalypse from below as a springboard for articulating apocalyptic eco-politics centred on the rejection of the world on normative grounds. However, the *ontology* operative in such a radical eco-apocalypticism would, unwittingly, stalk the fires of apocalypse from above – making it complicit in providing credence to top-down, authoritarian measures.

Disinvesting from the world

Thomas Lynch points that Taubes lacks "the sense of apocalypticism as an affectively charged response to the violence of the world." The point is not to announce one's spiritual disinvestment from the world but to learn "how to hate the world in me, engaging in a ceaseless process of disinvesting" (2022). This process – and the concomitant hatred of the world – begins when we realise that "the world itself is unethical":

> It is not the world, on top of which is laid capitalism, sexism, racism and other ideological formations. Those formations, in their complex

intersection, are the world. . . . These material and social relations cannot be resolved within the world, because they *are* the world. This impasse requires imaging the end of the world – a traumatic end that exceeds the legitimizing discourses of ethics and politics. . . . Such an end is the possibility of other possibilities.

(Lynch 2019, 30–32)

Lynch's intuitions are echoed by other discourses which reject and negate the world. In *No Future*, Lee Edelman argues against the attachment to futurity represented by the symbolic figure of the Child, advocating for the "negation of everything that would define itself, moralistically, as pro-life" (2004, 31). Nathalie Zaltzman notes how the slogan "Long live Death" was taken up by 19th- and 20th-century Spanish insurgents in their struggle against injustice, turning a death drive of revolution against the death drive of the status quo (in Malabou 2023). Similarly, Calvin L. Warren develops a black nihilist position which argues for "an *ontological revolution*, one that will destroy the world and its institutions." As Warren points out, "[s]uch thinking would lead us into an abyss. But we must face this abyss – its terror and majesty" (Warren 2018, 171–172). In the words of Elizabeth Pyne, discourses which reject the world in its entirety use radical negativity to confront "the negativity that drives the world's reproduction."

At issue, then, is a negative mode that would resist oppressive systems and renounce pretentions ever to purge the negativity that animates resistance; seek the end of the world that reproduces violence . . . and make visible how the possibility of possibility verges on collapse both in grasping the future and in rejecting it.

(2023, 48)

Here, a normative critique of our conditions of existence, accompanied by hatred of, and a process of disinvestment from the world, replaces the investment in the world characteristics of eco-ideologies. Nevertheless, and counter-intuitively, I would like to propose that such radical apocalypticism can be reconciled with climate politics, albeit in a qualified form. To substantiate this hypothesis, we must examine in more detail the ontology which Lynch proposes in addition to his affective-normative criticism of the world. Although Lynch's ontological commitments are initially very promising, they run into problems which call for a revision of his ontology.

Drawing on Catherine Malabou's work on plasticity, Lynch argues for the following three claims. First, Lynch maintains the contingency of being. The world can end because being itself is plastic, and, as such, it is characterised by "a perpetual process of transformation" (2019, 125). Ontological plasticity, therefore, means that everything could be otherwise, including the existence of the world. Second, plasticity possesses both a destructive and a creative power – plastic changes of being are characterised by a rearrangement of matter, a destruction of form and a giving of form (2019, 99). Consequently, the end of the world will lead to the emergence of new possibilities – apocalypse is "a plastic process of metamorphosis in which annihilation, explosion and emergence are joined in contradictory relation" (2019, 103). Third, Lynch refuses to propose positive visions of the post-apocalyptic future: the "question of what begins is beyond plastic apocalypticism" (2019, 103). Any plans for the emergent world would necessarily be limited by the possibilities of this (to-be-destroyed) world, undermining the radical break between worlds produced by the apocalypse.

At this stage, we can begin to glimpse how Lynch's project can speak to environmental goals. Hating the world and desiring its end would constitute a possibility, albeit cataclysmic, of the emergence of new forms of being. Ontological plasticity would ensure that the collapse of the world would create new possibilities of existence. In contrast to people like Bendell, however, for Lynch we can't anticipate or plan for the future to ensure the thriving of humanity; all we can hope for – if we can hope at all – is the resurrection of hope *after* the end of the world:

> Only after the hopes of this world have been abandoned and the world itself has fallen away, will it be possible to witness the emergence of new hope . . . the only hope is for the end and that hope is enough, for now.
> *(Lynch 2019, 136)*

To demonstrate the eco-political potential of Lynch's apocalypticism, we can read it alongside one of Malm's arguments. Malm put forward the idea that nature is autonomous – despite lacking agency, nature possesses a capacity to regulate its own behaviour outside of human control. Capitalism subdues and restrains the autonomy of nature to plunder it more effectively. Revolution, by liberating nature from the limits imposed on it by capitalism, would open up a space for the emergence of new environmental possibilities, developed autonomously by nature (Malm 2017). In contrast to his authoritarian eco-Leninism, Malm's "ecological autonomism" dovetails with Lynch's apocalyptic project. The autonomy of nature – its capacity to develop on its own outside of capitalism – is an embodiment of plastic being. Revolution (a concept belonging to the orbit of apocalypticism) would mark a certain end of the world, capable of unleashing the creative forces of nature, currently subdued by the mechanisms of global capitalism. Consequently, we should happily run the risk of destructively "clearing the space" to allow for a free (because unplanned and unexpected) development of the future.[1]

At this point, we can reformulate the category of apocalypse from below. If, as Taubes puts it, the katechon "holds down the chaos that pushes up from below" (2004, 103), then apocalypse *from below* would amount to the removal of the restrainer – a dictator, an empire, an economic system, or the world itself – resulting in the "anarchic unleashing" of the plastic being (Lynch 2019, 130). Moreover, contingency proper to ontological plasticity ensures both that the katechon can be toppled and that world-destruction opens a space for radical and unexpected transformations. Note, however, that this is also the ontology of apocalypse *from above*: the avowed function of the restrainer is to suppress and control the chaotic power of plasticity, which, if left to its own devices, could set the world on fire. The empire remembers that wild rivers and ungovernable jungles have consumed innumerable lives of colonial explores; the katechon, therefore, to protect itself, society, the economic order, or even the natural world (understood in reference to the katechon's interests), must subdue the autonomous forces of plastic being. The fact that apocalypses from below and above, despite their apparent opposition, hide an ontological agreement regarding the anarchic force of plastic being has two important consequences.

First, it suggests that being is on the side of apocalypse from below. The katechon wants to suppress the effects of unrestrained plasticity. Anti-katechonic politics, by contrast, would harness the power of, and align itself with, ontological plasticity, insofar as the latter is the condition of possibility of both the destruction of the katechon and the releasement of plastic being – "pushing up from below" – from its confines. Here we can notice the religious heritage of apocalypse from below: to believe that being "conspires" against its worldly prison, and that it has the power to abolish the katechon, is a materialist translation of God's investment in the history of salvation, which involves him destroying empires, nations, and worlds to liberate his people.

The second, more problematic consequence of the ontology shared by the two sides of the Taubesian distinction, is that apocalypse from below and apocalypse from above inform one another, creating a vicious cycle of apocalyptic (eco-)politics. The possibility of the world's end, grounded in the ontological plasticity of being, leads the katechon to introduce authoritarian measures whose goal is to prevent the catastrophe. These measures, in turn, are interpreted by some as unethical, generating a strong affective response and the concomitant need to disinvest from the world. Those who reject the world, in turn, formulate the demand for the end of the world, whose possibility is grounded in the plasticity of being. Inadvertently, this demand rearticulates apocalyptic threats for katechonic actors, which leads to the introduction of new preventative measures, postponing the end of the world. This vicious cycle is nothing other than an expression of the ideological function of apocalypticism explored in Chapter 1. The popular calls to end our current conditions of existence, often accompanied by a critique of specific forms of society's reproduction, identify the areas in which authorities can intervene, in effect neutralising the project of apocalypse from below.

It is my contention that it is possible to disengage the affective-normative rejection of the world from the ontology which perpetuates the vicious cycle of apocalyptic politics. To do so requires recognising that being *may not* change, that being is *not* on the side of apocalyptic revolutionaries, and that anti-world activism can have recourse to *both* restraining *and* releasing forms of apocalypticism. Paradoxically, to save the spirit of apocalypse from below, we should revise Lynch's ontology and side-step Taubes's categorical distinctions.

The law of contingency: too much, too little

Implicit in Lynch's ontology of plasticity is the distinction between the world and being. While the former ends, the latter ensure a creation of unanticipated possibilities. We can articulate this difference with the help of Emmanuel Levinas. In *Time and the Other*, Emmanuel Levinas carries out the following thought experiment: he asks his readers to imagine the annihilation of "all things, beings, and persons."

> What remains after this imaginary destruction of everything is not something, but the fact that there is [*il y a*]. The absence of everything returns as the presence . . . an atmospheric density, a plenitude of the void, or the murmur of silence. . . . The fact of existing imposes itself when there is no longer anything. And it is anonymous: there is neither anyone nor anything that takes this existence upon itself.
>
> *(1987, 46–47)*

The plastic being found in Lynch's ontology bears a striking resemblance to Levinas's *il y a*: it is what remains when the world ends, and as a remnant, it makes possible the appearance of new material configurations. Where Levinas and Lynch differ, however, is in the characteristics they find in being: Levinas stresses the oppressive inescapability of being, comparable to the frustrating experience of insomnia (Levinas ends up calling the *il y a* absurd and evil (1987, 50–51)); Lynch by contrast, by connecting being with plasticity, sees in it liberating contingency and destructive possibilities of novelty.

I believe that Levinas intuits something important about being, which Lynch misses: being is not only the liberating principle of change but also a principle of inescapable oppression. In fact, we can arrive at this conclusion by radicalising Lynch's own thesis regarding the contingency of being.

For Malabou, on whose work Lynch draws, plasticity is both the "law" of being and an adequate concept by which to grasp the ontological principle of change.

> [I]t should be clear that plasticity refers to both a new mode of being of form and a new grasp of this mode of being itself, in other words a

new *scheme . . . plasticity is the systemic law of the deconstructed real*, a mode of organization of the real that comes after metaphysics and that is appearing today in all the different domains of human activity . . . the privileged regime of change today is the continuous implosion of form, through which it recasts and reforms itself continually. Then also because we can only access these new organizations or configurations thanks to a tool that *conforms to these forms* itself, a tool that accords with them or is adequate to them.

(2010, 57)

Lynch takes up Malabou's characterisation of plasticity as an ontological law by affirming the *necessity of contingency*: insofar as everything is plastic, everything can be undone. Having said that, there is *one* thing that cannot be destroyed – the theoretical tools which track the ontological transformations. "Nothing is beyond change except the system of knowing which grasps the fundamental concepts inherent to that change" (Lynch 2019, 100). We can see what motivates Lynch to protect the notion of plasticity from the reach of ontological contingency: if the concept of plasticity is not necessary, it would be arbitrary; however, if plasticity is arbitrary, then it can no longer adequately name the law of being. As a result, the very idea of the plasticity of being could be put into question; we would end up with a radical view of contingency which Lynch rejects, where "it could all be different or not be at all" (2019, 122).

By embracing the necessity of contingency, therefore, Lynch shows himself to be invested in being as essentially *changeable*. However, this ontological commitment is problematic for two reasons. Plasticity, as the principle of being is the ontological condition of this and *any other world*. This means that apocalypse – understood as the end of this world's possibilities – is never absolute, because there is always at least one thing which *survives* the apocalypse: the ontological possibility of plastic change. While I personally don't think this is a problem, I don't believe that this is a conclusion which sits comfortably with Lynch's programmatic ban on anticipating future possibilities.

More significantly, however, the blind spot of Lynch's attachment to the necessity of contingency is that *lack of change* is ruled out in principle. Admittedly, observations of the world around shows that change is ubiquitous. Quentin Meillassoux calls this type of empirical contingency "precariousness," which "designates perishability that is bound to be realized sooner or later."

This book, this fruit, this man, this star, are all bound to perish sooner or later, so long as physical and organic laws remain as they have been up until now. Thus "precariousness" designates a possibility of not-being which must eventually be realised.

(Meillassoux 2009, 62)

Revolutions may end social and economic systems; climate change may transform our environmental conditions of existence; finally, apocalypses may annihilate worlds. The empirical evidence overwhelmingly points towards the necessary contingency of being and the concomitant change.

But, if *everything* is contingent, so should be the law of change. Meillassoux designates this fact as *absolute contingency*:

> [W]hat we see there is a rather menacing power – something insensible, and capable of destroying both things and worlds, of bringing forth monstrous absurdities, yet also of never doing anything . . . it is capable of destroying, without cause or reason, every physical law. . . . This is not a Heraclitean time, since it is not the eternal law of becoming, but rather the eternal and lawless possible becoming of every law. It is a Time capable of destroying even becoming itself by bringing forth, perhaps forever, fixity, stasis, and death.
>
> *(Meillassoux 2009, 64)*

Absolute contingency, therefore, can undo its own necessity. Change and transformations would constitute only one possibility of being; the other ontological modality would be one of persistence, rest, and stability. As Meillassoux puts it, absolute contingency destroys every law, which includes also the law of plasticity. Consequently, no theoretical categories are in principle necessary, especially not categories which grasp being as process of incessant transformation. In short, contingent being *may* or *may not* be plastic, and so it *may* or *may not* change. Interestingly, Lynch himself intuits this conclusion when engaging with Ernst Bloch ("Just because everything could be otherwise does not mean everything will be otherwise" (2019, 113)). However, the Blochian intuition is abandoned by Lynch, as it serves only to set up the discussion of necessity of plasticity in further sections of the book.

Fascinatingly, the ontological stability of being is experienced by climate denialists. "Everything trembles, but they don't, nor does the ground on which they stand. The framework in which their history unfolds is necessarily stable. The end of the world is only an idea" (Latour 2017, 207). Of course, climate denialists are *too* extreme, since they don't acknowledge that even the stable world *may* end. Yet their conviction can't be dismissed as simple ignorance or madness, since it is grounded in an ontologically justifiable experience, which accesses a possible modality of contingent being – its fixity.

The experience of being's absolute contingency can be illustrated with help of the film *Unrest* (2022). The story follows the young Pyotr Kropotkin during his visit to a Swiss town populated with anarchist watchmakers. As we learn, the titular unrest does not indicate revolutionary struggle with the powers that be (something we might have expected knowing that the main protagonist is the anarchist Kropotkin); rather, unrest refers to a spiral wheel, installed inside watches, which allows the mechanism of the clock to carry out its circular measurement of time and to control the time of the workers. Soon, it becomes apparent that both the Swiss village and the viewer are beholden to the mechanistic power of unrest. Although anarchists encounter nationalists and policemen, and the possibility of open struggle is permanently real, *nothing happens.* The film generates an experience of Levinasian *there is:* both the village and the film are characterised by an endless vigilance, a frustrating insomnia – an oppressive experience in which the anticipated end (of revolution or of sleep) never arrives. This experience is aptly captured by the title of Amy Nicholson's *New York Times* review of the film: "Times They Are Not A-Changing" (2023).

The notion of absolute contingency has direct consequences for anti-world politics attempting to preserve the spirit of apocalypse from below. First, the belief that plastic being "conspires" against the katechon can no longer be maintained. If being is radically contingent, it is structured by both change and persistence; contingency, therefore, is also the condition of the indefinite postponement of the apocalypse. As such, being *also* makes possible the work of the restrainer; ontology, just like Treebeard from *The Lord of the Rings,* is on nobody's side (Tolkien 1954).

Second, while apocalyptic eco-activist can remain motivated by an affective-normative rejection of the world, their strategy shouldn't be focused solely on the destruction of the world to "let things be" – simply because the apocalypse may result in *nothing* changing. Instead, to continue *The Lord of the Rings* analogy, apocalyptic eco-activists, to be effective, must "convince" being to join the struggle just like Merry and Pippin won Treebeard. This, in turn, requires a more complex strategy for the future, involving tactical engagement with multiple and ambivalent apocalypses and the employment of both releasing and restraining powers.

Bracketing the end of the world

The thesis regarding being's absolute contingency leads to a counter-intuitive conclusion from the point of view of apocalypticism: "we cannot claim to know for sure whether or not our world, although it is contingent, will actually come to an end one day" (Meillassoux 2009, 62). I would like to conclude this chapter by demonstrating how this claim, rather than neutralising apocalypticism, enables us to refocus it. In fact, "bracketing" the inevitability of end of the world can align apocalypticism which rejects the world on affective-normative grounds more effectively with environmental goals.

Let's return to our discussion of Maimonides in Chapter 1. We can understand Maimonides's treatment of apocalyptic prophecy as stemming from his seemingly incompatible philosophical and religious commitments.

Maimonides's *The Guide* notes that the "opinion of Aristotle is that the Universe, being permanent and indestructible, is also eternal and without beginning" (1956, 211). We can see how this philosophical view may conflict with Maimonides's religious beliefs. First, the Biblical account of creation presents the universe as having a beginning; second, the possibility of the destruction of the universe seems to be asserted by various passages found in the writings of the prophets.

Interestingly, in *The Guide*, it is only the first religious view which is defended by Maimonides against the philosophers – namely the belief that the universe was created out of nothing and that, therefore, it had a beginning (1956, 174–200). However, the second position, which might also seem to be religiously orthodox, is problematised. "It is not contrary to the tenets of our religion to assume that the Universe will continue to exist for ever" (1956, 201). As *The Guide* shows us by a careful Biblical exegesis, certain passages found across the books of the Bible seem to agree with the philosophical belief in the indestructibility of the universe (for instance, Eccles. 1:4 – "Generations come and go but the earth remains forever") (1956, 203). Consequently, "reasoning leads to the conclusion that the destruction of the Universe is not a certain fact" (1956, 202). A religious person, therefore, is free to take apocalyptic prophecies literally and assert that the universe will come to an end or to agree with the philosophers by accepting that the universe will last forever. Maimonides counts himself among those who agree with the philosophers with regard to the belief in the indestructibility of the universe. Nothing necessitates the conclusion that the universe will come to an end because "the existence or non-existence of things depends solely on the will of God and not on fixed law" (1956, 202). Eight hundred years before Meillassoux, Maimonides formulates the absolute contingency thesis, albeit in religious terms.

Importantly for our purposes, Maimonides shows how bracketing the necessity of the end of the world enables us to "splinter" the supposedly single apocalypse into multiple localisable apocalyptic events with political significance. As we have seen in Chapter 1, Maimonides reads apocalyptic prophecies as political allegories, treating of ends of specific societies. For instance, when Zephaniah prophesies that God will destroy everything earthly ("I will sweep everything from the face of the earth" Zeph. 1:2), he means "the ruin of a person, of a nation, or of a country" (1956, 204) and not the destruction of the world. Maimonides, therefore, can maintain both that the world may or may not end and that we are surrounded by multiple, local, and politicised apocalypses.

The theoretical tension experienced by Maimonides – between the Biblical belief in the end of the world and the Aristotelian commitment to the perpetuity of the universe – in a certain a way mirrors the paradox which structures ecological apocalypticism. How can we both desire the end of the world and be concerned with our environmental conditions of existence?

The solution offered by Maimonides will be of help in indicating an approach to this paradox, which I will develop in the following chapters. Bracketing the inevitability of the end of the world allows us to splinter the climate apocalypse and to represent the climate crisis as populated with many differentiated eco-apocalypses. The starting point of climate apocalypticism, therefore, is a web of localisable "ends of the world," with distinct political meanings. We can draw on the conceptual orbit of apocalypticism to name these ends, to specify their character, and to devise appropriate tactics of engagement.

Consequently, the affective-normative negation of the world, while still operative as a guiding principle, would have to be qualified in the case of each apocalypse. Some ends must be hastened, others initiated, and others still – *stopped*. In other words, to remain faithful to the injunction to let unethical worlds go down, sometimes we must play the role of the katechon. The analogy here can be found in the early proletarian campaign for the reduction of the working day – as Jason Read notes, withholding work "not only liberated the time of the workers; it also profoundly transformed production" (2024, 144).

In other words, the only way to effectively divest from the world is to harness *both* the releasing *and* withholding apocalypticism to influence the planetary (non-)reproduction of our conditions of existence. While such an approach blurs the Taubesian distinction between apocalypse from below and apocalypse from above, it opens up a possibility for anti-world climate politics to strategically play good, bad, and neutral aspects of apocalypses against each other, enabling it to gain an upper hand in the inevitable confrontation with its political enemies – or so we can hope.

Note

1 Interestingly, the Catholic Church in Ireland has decided to return 30% of church ground to nature (Kurian 2023). While this initiative is certainly less dramatic than the passage through revolution or the end of the world, it operates with the same ontology of releasing the autonomy of natural being by "clearing the space" for it.

Bibliography

Althusser, L. (2011) *Philosophy and the Spontaneous Philosophy of the Scientists.* Trans. B. Brewster et al. London: Verso

Bendell, J. (2020) "Adapting Deeply to Likely Collapse: An Enhanced Agenda for Climate Activists?" Available online: https://jembendell.com/2020/01/15/adapting-deeply-to-likely-collapse-an-enhanced-agenda-for-climate-activists/ [Accessed 26/10/2024]

Curtis, N. (2022) "Red Pill: The Structure of Contemporary Apocalypse," *Apocalyptica* 1(1), pp. 96–116

Edelman, L. (2004) *No Future: Queer Theory and the Death Drive*. Durham, NC: Duke University Press

Gittlitz, A.M. (2020) *I Want to Believe: Posadism, UFOs and Apocalypse Communism*. London: Pluto Press

The Holy Bible. (2011) *New International Version*. Palmer Lake: Biblica

The Invisible Committee. (2015) *To Our Friends*. Trans. R. Hurley. Los Angeles: Semiotext(e)

Kurian, A. (2023) "Ireland's Bishops Decide to Return 30% of Church Grounds to Nature by 2030," *Catholic News Agency*. Available online: www.catholicnews agency.com/news/255165/ireland-s-bishops-decide-to-return-30-percent-of-church-grounds-to-nature-by-2030 [Accessed 26/10/2024]

Latour, B. (2017) *Facing Gaia: Eight Lectures on the New Climatic Regime*. Trans. C. Porter. Cambridge: Polity Press

Levinas, E. (1987) *Time and the Other*. Trans. R.A. Cohen. Pittsburgh: Duquesne University Press

Lynch, T. (2019) *Apocalyptic Political Theology: Hegel, Taubes and Malabou*. London: Bloomsbury

Lynch, T. (2022) "How I Learned to Stop Hoping and Hate the World," Available online: https://cursor.pubpub.org/pub/issue8-lynch-how-i-learned/release/1 [Accessed 19/10/2024]

Maimonides, M. (1956) *The Guide for the Preplexed*. Trans. M. Friedländer. New York: Dover Publications, Inc.

Malabou, C. (2010) *Plasticity at the Dusk of Writing: Dialectic, Destruction, Deconstruction*. Trans. C. Shread. New York: Columbia University Press

Malabou, C. (2023) "Being an Anarchist," *Ill Will*. Available online: https://illwill. com/being-an-anarchist [Accessed 26/10/2024]

Malm, A. (2017) *The Progress of This Storm: Nature and Society in a Warming World*. London: Verso

Malm, A. (2020a) *Corona, Climate, Chronic Emergency: War Communism in the Twenty-First Century*. London: Verso

Malm, A. (2020b) *How to Blow Up a Pipeline: Learning to Fight in a World on Fire*. London: Verso

Meillassoux, Q. (2009) *After Finitude: An Essay on the Necessity of Contingency*. Trans. R. Brassier. London: Bloomsbury Publishing

Nicholson, A. (2023) "Unrest' Review: The Times Are Not A-Changin," *The New York Times*. Available online: www.nytimes.com/2023/05/04/movies/unrest-review.html [Accessed 26/10/2024]

Pyne, E. (2023) "Queer Ecologies and Apocalyptic Thinking," in *The Environmental Apocalypse: Interdisciplinary Reflections on the Climate Crisis*, J. Kowalewski (Ed.). Abingdon: Routledge

Read, J. (2024) *The Double Shift: Spinoza and Marx on the Politics of Work*. London: Verso

Schmitt, C. (2006) *The Nomos of the Earth in the International Law of the Jus Publicum Europaeum*. Trans. G.L. Ulmen. New York: Telos Press Publishing

Skrimshire, S. (2019) "Activism for End Times: Millenarian Belief in an Age of Climate Emergency," *Political Theology* 20(6), pp. 518–536

Taubes, J. (2004) *The Political Theology of Paul*. Trans. D. Hollander. Stanford: Stanford University Press

Taubes, J. (2013) *To Carl Schmitt: Letters and Reflections*. Trans. K. Tribe. New York: Columbia University Press

Tolkien, J.R.R. (1954) *The Two Towers*. London: George Allen & Unwin

Unrest. (2022) Dir. Cyril Schäublin. Switzerland: Cinédokké

Warren, C.L. (2018) *Ontological Terror: Blackness, Nihilism, and Emancipation*. Durham, NC: Duke University Press

4

THE SHAPES OF ECO-APOCALYPTIC TIME*

Delf Rothe has argued that contemporary Western responses to the climate crisis are "deeply influenced by a linear temporality and a common orientation towards the threat of the end of time" derived from Christian eschatology (2020, 145). The various genres of environmental politics share one presupposition – a belief in time as a single, unidirectional line tending towards climate apocalypse, which mirrors the Christian view of history as "a flow or movement from a starting point (the creation) towards a final event in the divine plan (the eschaton)" (2020, 156). This leads Rothe to conclude that "the discourse on the Anthropocene is essentially eschatological" (2020, 147).

However, the eschatological belief that historical time is a single line leading to an apocalyptic event generates two serious, interrelated problems for any environmentalism, which, together, constitute what I call a decolonial critique of eco-apocalypticism:

P1: The linear view of time centred around a present climate crisis or a future ecological catastrophe "disregards that many people in the majority world have already lived through the ecological catastrophe brought about by European colonialism and its repercussions" (Rothe 2020, 146). For example, as Déborah Danowski and Eduardo Viveiros de Castro point out, "for the native people of the Americas, *the end of the world already happened* – five centuries ago" (2016, 104). However, the exclusive focus on

* The earlier version of this chapter was previously published as Kowalewski, J. (2023) "The Shapes of Apocalyptic Time: Decolonising Eco-Eschatology," in *The Environmental Apocalypse: Interdisciplinary Reflections on the Climate Crisis*, J. Kowalewski (Ed.). Abingdon: Routledge. The current version of the chapter has been updated and expanded.

DOI: 10.4324/9781003348511-5

the environmental disaster as a future or present end of linear history blinds us to ends of history experienced by colonised communities in the past.

P2: The single timeline, expressed for instance in a narrative about future human extinction common to contemporary eco-apocalyptic discourses, gives the impression that "we are all in this together" (Rothe 2020, 146). This, in turn, de-politicises the environmental emergency by obscuring the historically and geographically specific effects of climate change. As Anupama Ranawana and James Trafford put it, the faux "anti-political universalism" of a single apocalyptic narrative "actively conceals both how climate crises are temporally and spatially distributed, and how they are symptoms of ongoing imperialist practices" (2019).

One of the aims of this chapter is to suggest that climate apocalypticism does not *necessarily* reproduce the Western-centric standpoint which leads to P1 and P2. In fact, as Danowski and Viveiros de Castro make clear in their work, the eschatological focus on the "end of the world" can effectively lend itself to a decolonial project. Eco-eschatologies, therefore, don't have to be abandoned altogether, although their understanding of time and space must be rethought in light of the decolonial critique presented earlier. This chapter, therefore, offers a theoretical corrective to eco-apocalyptic temporality by proposing an alternative model of eschatological time capable of addressing both P1 and P2.

First, I will argue that the model of historical time found in apocalyptic literature is not a line but a *spiral* which combines linear and cyclical elements. Such an understanding of time would respond to P1 by recognising the connection between the past, present, and future apocalypses and the constitutive role of past ends of the world for an eschatological history. Second, I will demonstrate that apocalyptic discourse presupposes multiple timelines, whose relationship can be understood as a dislocated and non-contemporaneous historical totality. I will sketch the latter with the help of Louis Althusser to show how such a model of time can address P2. I will conclude this chapter by suggesting that a twofold understanding of apocalyptic time – as a spiral and as a non-contemporaneous totality – can help us to devise, respectively, tactics and strategy for eco-apocalyptic politics.

A line or a cycle?

In her book on apocalypse in Japanese science fiction, Motoko Tanaka rightly observes that a lot of "research on world eschatologies refers to the major distinction between linear historical apocalypses and cyclical traditional apocalypses." In contrast to the popular characterisation of the Judeo-Christian apocalypse as a final event of a linear history,

cyclical apocalyptic narratives envision no absolute end; many of these stories and myths envision apocalyptic destruction at the end of each cycle, yet assume that the world will be restored in the next rotation. Cyclical eschatology has its end, but implicit in this ending are both restoration and rebirth.

(Tanaka 2014, 16)

The distinction between these two models of time might initially lead us to substitute the linear conception of history with a single end (rejected on the basis of the decolonial critique) with the cyclical image of historical time with multiple ends. However, I believe that the cyclical time cannot offer a viable eco-apocalyptic alternative to linear time.

If we conceive of historical time as a cycle, we are led to two contradictory conclusions regarding the end of the world. First, there is no singular apocalypse which fulfils time; rather, history's end repeats itself at and as the end of each cycle, generating a *multiplicity* of the world's ends. Second, and in contrast to our first claim, if history is a cycle, it must be *the same end* which is repeated each time a cycle reaches its conclusion. If this wasn't the case, and *end X* and *end Y* were, in fact, qualitatively distinct, then we could present the relationship between these two ends as a linear process *from* end X *to* end Y – thus abandoning the commitment to a purely cyclical model of time. Consequently, the multiplicity of the world's ends generated by strictly cyclical history is only *apparent* – in fact, it is the same, single end which has taken place in the past and will take place in the future. As such, cyclical eschatology repeats the belief in a single end characteristic of its linear counterpart.

Furthermore, when considered from an existential point of view, cyclical history can generate a need to aim for a qualitatively distinct future. As Mircea Eliade puts it in his study on the eternal return in religions:

[A]ll these numberless aeons also have a soteriological function; simply contemplating the panorama of them terrifies man and forces him to realize that he must begin this same transitory existence and endure the same endless sufferings over again, millions upon millions of times; this results in intensifying his will to escape, that is, in impelling him to transcend his condition of "living being" once and for all.

(1959, 115–116)

However, to truly escape cyclical history, and to transform our existential condition "once and for all," the cycle must be broken, and time must become a line, since it is only a linear model of time which can make sense of an *irreversible* change between two distinct temporal points necessary for an existential "escape." Thus, even if history is cyclical, it makes a subjective demand for an end conceived on the basis of a linear temporality.

From an eco-apocalyptic perspective, therefore, cyclical history is not a feasible replacement for linear time. First, on the cyclical view of time, we can acknowledge past ends of the world only by equating them with present and future apocalypses – thus creating a narrative of an eternally returning extinction event. Such a narrative, however, cannot capture the temporal and spatial *differentiation* of the apocalypses. In other words, although the cyclical model of history can account for past ends of the world (thus addressing P1), it can do so only by reducing the *specific differences* between the multiple ends and thus running into P2, that is, the faux universalism of a single extinction story centred on *the same*, recurring apocalypse. Second, the existential demand to overcome cyclical time refocuses our attention on a qualitatively distinct end which has to be located in the future – which, in turn, risks reintroducing P1 with its blindness to past ends of the world.

A spiral

Since both purely linear and purely cyclical models of eco-apocalyptic time are flawed, in what follows I will suggest a third shape of time: a *spiral* which combines linear and cyclical elements. Such a model of historical time can be reconstructed on the basis of medieval and contemporary apocalyptic literature. In my reconstruction, I will first draw on the works of 12th-century monk Joachim of Fiore and then turn to 20th-century philosopher Jacob Taubes.[1]

Joachim is perhaps best known for his division of history into three stages or *status*: the age of the Father – corresponding to the events of the Old Testament; the age of the Son – marked by the domination of the Catholic Church; and the final age of the Holy Spirit – which Joachim envisaged as a time of monastic orders.[2]

The division of history into three stages has been highly influential. As Jacob Taubes points out, "the model of antiquity – Middle Ages – modern age is nothing but a secular extension of Joachim's prophecy of the three ages of the Father, the Son, and the Holy Spirit" (2009, 82). Also, we can map the Marxist periodisation of history into feudalism, capitalism, and communism, as well as Comte's theological, metaphysical, and positive stages of society onto the three *status* of Joachim. Interestingly, some decolonial thinkers also employ categories and theoretical models which possess an undoubtable Joachimite inflection. For example, in *The Politics of Decolonial Investigation*, Walter D. Mignolo argues – in a Joachimite manner – that "we, on the planet, are experiencing a change of epochs": a movement from the "second nomos" understood as "unipolarity in international relations and the hegemony/dominance of Western modernity" and the "third nomos" marked by "de-Westernization and decoloniality" (Mignolo 2021, 484–485).

Joachim frames the events of the apocalypse as a transition from the era of the Father, through the era of the Son, to the era of the Holy Spirit. Although the progressive movement through the three eras is undoubtedly built on a linear model of time, in Joachim's conception of history the linearity of history is often articulated in terms of cycles.

Joachim, a prolific illustrator, famously pictures historical time as three intertwined circles. Bernard McGinn describes Joachim's drawing in the following way:

> Three interlocking rings demonstrate how the mystery of the Trinity relates to the course of history. The first circle in green belongs to the Father and forms the time of the Old Testament. The middle blue circle, that of the Son, interlocks with both extremities – the median that joins the extremes. The final flaming red circle of the Holy Spirit indicates the double procession of the Third Person by intersection with both the green and the blue circle . . . the noninterlocking area of this last circle does suggest a coming special era within history, not unrelated, but still superior to what has gone before.
>
> *(1979b, 104)*[3]

One of the possible reasons why Joachim thinks of the relationship between the three stages of history as interlocking circles is that he views history as structured around a repetition of meaning – or *concordance* – found in the parallels between people and events of the Old and the New Testaments.[4] For Joachim, history moves forward in a cyclical movement which repeats the significance of particular characters and events across the past and the present *status*. As Marjorie Reeves explains:

> The medieval approach to history which sought in each episode an inner meaning which linked it by concord with events of another era is, of course, quite foreign to us. It is as if each happening had a vertical point of reference, a "thread" in the hand of God who combined threads into patters on the inner side of history, whereas we look only for the horizontal connections and the pattern of visible cause and effect spun along the time span. . . . Thus the three chief patriarchs, Abraham, Isaac and Jacob, are in concord with Zacharias, John the Baptist and Christ, and, of course, typify the three Persons of the Trinity and, in consequence, the three *status* of history.
>
> *(1999, 10–11)*

What is the goal of constructing concordances? For Joachim, the concordance between characters and events of the present and past *status* offers a possibility of anticipating the shape of the future era. In other words, Joachim shows how apocalyptic reading can offer an insight into historical tendencies capable of delimiting future possibilities.

The notion of concordance, therefore, cannot reduce the differences between given characters or events, merging them into one, single entity. If it did so, history would be purely circular – since it would be constituted by the repetition of the same event or character – and consequently could not be moving towards the next, qualitatively distinct *status*. Concordance finds identities across history; however, what the parallels reveal are specific historical developments between repeated meanings, significant from the point of view of the era to come, because it is these developments (say, between Abraham, Isaac and Jacob, and Zacharias, John the Baptist and Christ) which allow us to anticipate the next *status*. In the words of McGinn, "These letter-to-letter comparison and parallels between the Testaments are not used merely to understand the past, but also, and far more daringly, to reveal the future. . . . The Old Testament, the New Testament, and especially the Book of the Apocalypse, when illuminated by the typological understanding, can show the meaning of what is to come" (McGinn 1979b, 102). Thus, for Joachim, eschatological predictions presuppose a *spiral* movement of historical time, where historical meanings move *forward cyclically*.

In response to a criticism that Joachim doesn't work with history but with a Biblical material, and so concordances are at best a *literary construction* of history, we can note that concordances have also been used to discuss non-Biblical events and characters. For instance, Karl Marx writes in *The Eighteenth Brumaire of Louis Bonaparte* that "the Revolution of 1789 to1814 draped itself alternately as the Roman republic and the Roman empire, and the Revolution of 1848 knew nothing better to do than to parody, now 1789, now the revolutionary tradition of 1793 to 95" (1972, 10).

The spiral shape of history, constructed around concordances, can also be found in Taubes's philosophical history of apocalypticism, published in 1947 as *Occidental Eschatology*. Taubes's text seems to be characterised by an apparent tension between a commitment to a linear view of time and the recognition of the cyclical return of the apocalyptic theme throughout Western intellectual history. On the one hand, Taubes – a self-proclaimed apocalyptic (2004, 103) – states explicitly that for apocalypticism "History does not complete a circle" (2009, 33) and that, instead, "In the once-was of creation history has its beginning, and in the one-day of redemption it comes to its end. The interim between creation and redemption is the pathway of history" (2009, 13). However, this seeming philosophical commitment to a linear time is contrasted in the course of Taubes's exposition by a repetition of apocalypticism as a historical phenomenon – as Taubes makes clear, the expectation of the end-times is a motif which returns cyclically, from antiquity to modernity. It is precisely this repetition of apocalyptic theories and experiences which allows Taubes to construct concordances, for example, between ancient apostles and modern philosophers.[5]

The apparent tension in *Occidental Eschatology* between the linear and cyclical aspects of historical time can be easily resolved if we emphasise that

Taubes is committed to a *spiral* thinking about both apocalypticism and history.[6] On the spiral model, the apocalyptic motif is both repeated and altered throughout the forward march of time. Furthermore – in a manner reminiscent of Joachim – Taubes uses the cyclical parallels between various figures in the history of apocalypticism, and the developments between them, to predict a new era in the history of eschatology (admittedly, in rather obscure terms).[7] Taubes, therefore, seems "in concord" with Joachim in at least three ways: (1) they are both apocalyptic thinkers interested in the history of the end of the world; (2) they both emphasise the repetition and development of specific meanings throughout this history, which demonstrates their commitment to the spiral character of time; (3) for both Taubes and Joachim, the analysis of the spirality of historical time serves to anticipate the next *status*.[8] Insofar as making predictions "is one of the essential purposes of science: astronomers forecast the next lunar eclipse, meteorologists tomorrow's weather, economists movements in share prices, geologists the threat of earthquakes" (Dupuy 2022, 108), the apocalypticisms of Joachim and Taubes would constitute a genuine case of eschatology in the etymological sense: the *science* of the last things.

The status of concordances

One may object that there is a significant difference between apocalyptic and scientific predictions. Joachim and Taubes project their own interests, and the concomitant meanings, onto historical material; in other words, although they may think they *find* parallels in history, they in fact *create* them. In consequence, while such repetitions may be of interest to intellectual historians, they can't help us anticipate the future era: as *subjective* creations, concordances can't track objective historical developments. In short, the critic concludes, concordances, in contrast to scientific predictions proper, are simply arbitrary conjectures. This objection finds a surprising support in the writings of Freud.

When one encounters the number 62 multiple times in the same day – for example, one sees it on a ticket, on the train one boards, or on a door of a hotel room – one can be excused for according it "a secret significance to the persistent recurrence of this one number – to see it, for instance, as a pointer to his allotted life-span" (Freud 2003, 145). As Freud points out, stumbling upon a repetition can be uncanny, leading us to believe the world is trying to share with us a hidden meaning. Of course, in reality, it is *me* who notices the numbers and who unconsciously constructs both the repetitive patterns and their secret meaning.

The uncanny encounter with repetition, wholly produced by us, is even more visible in Freud's own experience of returning to the same area while wandering aimlessly in a foreign town. This area, we should note, is inhabited by sex workers; when Freud realises it, he "hastily left the narrow street at the next turning."

> However, after wandering about for some time without asking the way, I suddenly found myself back in the same street, where my presence began to attract attention. Once more I hurried away, only to return there again by a different route. I was now seized by a feeling that I can only describe as uncanny, and I was glad to find my way back to the piazza that I had recently left and refrain from any further voyages of discovery.
>
> *(Freud 2003, 144)*

The examples given by Freud – of seeing patterns, finding meaning, and returning to the same place – show that the subject is complicit in the construction of different forms of repetition. It also shows that this complicitly is covered over by an uncanny effect of discovery, as if we had stumbled upon an externally existent pattern or direction. Following Freud, it can be suggested that concordances – though they may seem like independently existent historical parallels and tendencies – are simply subjective creations, more akin to literary association than to scientific prediction.

However, rather than undermining concordances, I believe that the Freudian criticism can help us foreground the importance of the subject in the study of historical parallels and to understand them to be reflective of the subjective position of the author. Let's accept that concordances are constructed, and this construction is the work of the subject. But the subject doesn't exist in a vacuum; rather, she works from *within* history *with* historical material. The number 62 is in fact found in the world; and the area which Freud supposedly wants to avoid is an actual district of an actually existing town. Similarly, repetitions in history are constructed *out of* sources, characters, or events. Thus, concordances are the creation of the subject who, steeped in history, finds initially illegible connections and who endows these connections with meaning. The uncanny effect of such a work may lead some authors to fall into a trap of believing in a "discovery," obscuring their role in the appearance of historical parallels.

To take Joachim as an example: he started his career as a Benedictine monk, continuing to obey the Rule of St Benedict throughout his life (McGinn 1979a, 126). Consequently, and unsurprisingly for a reader familiar with Joachim's biography, he links the beginning of the final status of history to the appearance of St Benedict (in McGinn 1979a, 134). By contrast, the Franciscan Peter Olivi takes up Joachim's model of history but substitutes St Francis for St Benedict when marking the eschatological transition to a new era. A similar role of subject positions can be found in more recent use of concordances: when identifying historical repetitions, Marx takes "the pulse of history . . . listening to a revolutionary *frequency*" (Derrida 1994, 140).

The revolutionary bias of Marx can be contrasted with Derrida, who, as we have seen in this book's Introduction, situates the repetition of the "end of history" in the context of the (post-)Cold War anxieties of his time.

The work of the subject is necessitated by the fact that, as Althusser points out, historical facts – even those which cause "mutation in the existing structural relation" – are often repressed, purposefully or unconsciously hidden from view (Althusser et al. 2015, 249–50). Note that this is also the belief operative in the decolonial critique; the very problem with the model of history found in Western eco-apocalypticism lies in obscuring ends of the world which affected and continue to affect colonialised communities. This covering over, in turn, produces a subjective demand to construct a model of history capable of unveiling hidden historical material.

While the fact that historical tendencies, to become legible, must pass through subjectivity doesn't undermine concordances and the predictions they make possible, the constitutive role of the subject also doesn't accord every predictive concordance equal value. As Dupuy observes, "human beings do not make predictions solely in order to know the future; they make them also in order to act upon the world" (2022, 108). Following Dupuy, we can suggest that one of the criteria by which to judge concordances is the extent to which they make possible effective political interventions.

Reaction and prevention

The focus on Joachim – for whom history involves a progressive movement through the *status*, culminating in the era of the Holy Spirit – may give a wrong impression that the spiral model of historical time *only* lends itself to progressive and emancipatory kinds of apocalypticism. Furthermore, the fact that Taubes can be seen as an heir to Joachim can, again mistakenly, suggest that spiral history implies apocalypse from below examined in the preceding chapter, that is, a standpoint which *welcomes* the end of the world. The reality is more complex: concordances and the spiral shape of historical time can be employed by any apocalyptic project, including *reactionary* and *preventative* ones.

Let's take Heidegger as an example. As Richard Wolin argues, for Heidegger, the only way to "shatter the 2,000-year reign of [Western] metaphysics" is to expose everything "to complete and total devastation" (Wolin 2023, 60).

> Heidegger was quick to add that this anticipated catastrophe should neither be "feared" nor construed as a "misfortune." Instead, he insisted that humanity should *welcome* this act of "self-obliteration," since it portended the "purification of Being from the thoroughgoing distortions resulting from the predominance of 'beings.'"
>
> *(Wolin 2023, 67)*

As those familiar with Heidegger's biography might have suspected, this apocalyptic ontology had a political underside: National Socialism. In fact, the very notion of "purification of Being" "paralleled National Socialism's obsession with racial purity" and the " 'cleansing fantasies' of Nazi race doctrine" (Wolin 2023, 67–68). Writing in 1943, Heidegger states:

> Only the Germans – and they alone – can redeem the West and its history. . . . The planet is in flames. The essence of man is out of joint. Only from the Germans, provided that they discover and preserve their Germanness, can world-historical consciousness arrive.
>
> *(in Wolin 2023, 60)*

Importantly, as Wolin notes, the eschatological notion of "Germanness" was for Heidegger grounded in "his commitment to National Socialism as a reenactment of the 'Greek Beginning': a 'repetition' that took place on a higher ontological plane" (Wolin 2023, 60). It is this repetition which structures the belief that the fires of World War II, combined with the destruction of metaphysics, mark a transition towards a new historical era.

Heidegger's reactionary apocalypse provides an important argument in support of preventative apocalypticism – some apocalypses, such as the one awaited by the German philosopher must be stopped. Importantly, the work of prevention can be done with the same tools of historical spirals and concordances.

Recently, Dupuy developed a type of precautionary apocalypticism where, to avert a future catastrophe, we must treat it as necessary – "the prospect of catastrophe can be made credible only if we can be persuaded first of its reality . . . the predicted sufferings and deaths will inevitably occur." The goal is, first, to project ourselves into the necessary future and, second, to reintegrate the "memory" of the coming catastrophe into our present "conceiving of the event in the future perfect tense . . . it *will have* taken place" (Dupuy 2015, 8). The productive paradox which Dupuy exploits is that by acting on the "memory of the future," we can avert the necessary apocalypse. The anticipated catastrophe, therefore, must be treated as "both necessary and improbable" (Dupuy 2015, 9).

By placing the future in the past to orient the present – with the goal of ultimately changing the future, Dupuy employs a spiral model of time. However, he does so in a modified way: rather than examining past repetitions to anticipate the future era, Dupuy *starts* with the future world to abort its arrival. However, it is clear that "the memory of the future" can be solidified by memories of the past – if the necessary future resembles or repeats a catastrophic past (say, a past of war, nationalism, and genocide), that is if the future is in concord with the past, both the necessity of the catastrophe and the need to avert are reinforced.

From P1 to P2

At this point we are equipped to answer P1. As you recall, the first decolonial objection to eco-apocalypticism accused the latter of being concerned exclusively with the present climate crisis or a future ecological catastrophe, consequently disregarding and covering over past "ends of the world" – suffered, for example, by colonised communities.

In contrast to a purely linear time with its exclusive interest in the present or future ends, the spiral shape of history enables us to account for past ends of the worlds in its concordance with ends-to-come. In other words, present and future climate disasters should be understood as related to and oriented by ecological catastrophes of the past. In fact, the attention to concordances is already implicit in many discussions on apocalypse. As Omar Rafeal Regalado Fernandez observes in relation to ecological and nuclear catastrophes:

> The mass extinction at the end of the Cretaceous, 66 million years ago, that led to the extinction of many animals like dinosaurs, pterosaurs, and marine reptiles has become a symbol of what an extinction-level event is. Towards the last years of the Cold War, the only event on Earth powerful enough to cause human extinction was a nuclear fallout. In the 1980s, it was discovered that a meteorite triggered an ecological collapse, and the collision of a meteorite was likened to a nuclear winter, consolidating them as symbols of the end of times.
>
> *(Regalado Fernandez 2023, 3)*

Moreover, and in contrast to a purely cyclical time, we are not condemned to continually repeating the same end. On the contrary, the parallels between the worlds' ends can help us to identify historical developments and thus to illuminate ways to "break the cycle" either by preventing catastrophes or by anticipating a qualitatively distinct *status* demanded by our existential situation. For example, past extinction events can be seen as offering insights into the character and the conditions of future extinction events; but in so doing, they also point towards changes necessary for a habitable earth for communities most affected by the destruction of their ecosystems. In short, the spiral model of history allows us to address P1 because it views past apocalypses as necessary both for understanding present and future environmental catastrophes and for putting forward transformative political solutions.

However, our answer to P1 encounters two related problems. First, as long as we conceive of history as single timeline – even if this timeline is spiral – we face P2, that is the problem of universal extinction narrative which depoliticises the environmental disaster by concealing the unequal distribution of the climate crises. Second, the single spiral timeline is also inadequate to account for the multiplicity of apocalypses, with diverse normative characters. Some apocalypses must be stopped, others can be brought about; furthermore, there are different ways of both preventing and hastening the end of the world. There is a world of difference between Taubesian and Heideggerian apocalypticisms; similarly, the katechon in Dupuy and Schmitt differs. The single spiral timeline, however, cannot track the many distinct apocalypses and apocalypticisms and the political tactics of *simultaneously* bringing about some and stopping other ends of the world.

To address these problems, I will develop a model of historical time capable of capturing the differentiated relation between multiple ends of the world. Moreover, as I will argue, such a model, when combined with the spiral view of history, can inform both tactical and strategic responses to climate apocalypse.

Four objects, four temporalities

To make repressed parts of history visible, Althusser suggests, we should "construct the concepts of the different historical times . . . out of the differential nature and differential articulation of their objects in the structure of the whole" (Althusser et al. 2015, 250). In the case of economy, for example, historical time "must be constructed out of the reality of the different rhythms which punctuate the different operations of production, circulation, and distribution" (Althusser et al. 2015, 248). If we follow Althusser, we can ask: what are the "objects" proper to the field of apocalypticism, and what temporalities do they presuppose?

The first object is *lived experience* – the experience of apocalyptic prophets produces in them a subjective certainty that they live in the end-times (Taubes 2009, 32); the grasp of time as a cycle of endless suffering may result in an experience of terror and a will to escape. Consequently, we can posit the existence of an *existential timeline*, an apocalyptic temporality which pertains to subjective experience. Since this "level" is constituted by individual perspectives, it is essentially multiple.

The second object – or rather, objects – are constituted by *historical concordances*, for example, between the Old and the New Testaments (in the case of Joachim), between historical events (in the case of Marx), and between textual sources and history (in the case of Taubes). Here, apocalyptic predictions stem from the analysis of historical tendencies. Thus, we can posit a second temporal "level": *historiographical time* concerned with the multiple repetitions and parallels found across history.

The third and fourth objects are, respectively, *planetary* and *cosmic* catastrophes. Here we find the *geological timeline* – concerned with process and events which extend beyond human lifespan, or even the lifespan of generations, and which include major extinction events which predate humanity

(Brannen 2017); and a *cosmological timeline* – found in religious beliefs about the destruction and (possible) rebirth of the universe or in theoretical explorations of the consequences of the eventual solar catastrophe (Brassier 2007). Although both apocalypses assume an extra-human perspective, the cosmological timeline also presupposes an extraterrestrial outlook.

Of course, these diverse temporalities are connected, co-constituting a structural whole. In the next sections I will explore how we should represent their interconnection. However, first, let me briefly discuss two benefits of distinguishing between different apocalyptic timelines.

As Joanna Zylinska notes, in recent years the possibility of human extinction – exacerbated by the threat of planetary climate change – has led to resurfacing of fantasies about interplanetary travel. The response to earthly apocalypse is to colonise other planets – a project made possible and led by (male) technological "geniuses" (Zylinska 2018). We can begin to see how the focus on differentiated timelines can provide a starting point for a critique of the male superhero fantasy, which, when re-articulated in terms of the relation between multiple times, becomes even less feasible. This gendered fantasy involves a subordination of existential and geological times to the time of selected individuals, somehow capable of exploiting cosmological time – a daring, if not outright ridiculous, suggestion.

Furthermore, the existence of the four levels of historical time pertinent to apocalypticism means that we can bracket any of the timelines, or focus solely on only one time, and still firmly remain with the orbit of apocalypticism. For example, leaving the question of the eventual destruction of the universe open (as we did in at the end of the preceding chapter) doesn't neutralise the problem of the apocalypse; on the contrary, it refocuses it on the remaining geological, historiographical, and existential timelines. Conversely, (as I will demonstrate in further chapters), a close analysis of lived experience is not tantamount to obscuring the wider problem of collective or planetary ends of the world; rather, it offers access to a distinct time of a discrete apocalyptic object.

Sediments of time and the (non-)contemporaneity of the whole

How should we represent the structural relationship between the different times of the apocalypse? For the sake of clarity – and to foreground its political potential – in my answer I will focus on existential, historiographical, geological times; however, the model I will propose can also be extended to the cosmological timeline.

On one reading, which I believe is erroneous, existential time would be situated within historiographical time while both would be grounded in geological time. The difference between the levels would be constituted by the *scale* of each temporality – from subjective, through historiographical, to geological. On this reading, the three levels could be visualised as gradually expanding circles of a *continuous* and *contemporaneous* temporal space in the shape of a cone, with existential time on top, historiographical time in the middle, and geological time at the bottom.

A similar model is suggested by Reinhard Koselleck's metaphor of the sediments or layers of time. The first layer is constituted by singular and irreversible events, important moments for personal and collective identity (e.g. Saul becoming Paul, revolution of 1789). These singular instances are themselves based on the second layer of sedimented time: the "structures of repetition" – long-lasting historical processes that make possible the emergence of an event, which themselves can undergo change, albeit a slow one. Finally, Koselleck discusses the third layer of historical time, which points "beyond the experience of individuals and generations." Here the examples are "biologically conditioned reproduction" and trans-generational cultural artifacts, including "the realm of religious or metaphysical truths" (Koselleck 2018, 8–9). Although each level possesses distinct rhythms and velocities, each layer is placed "on top" of the other – as the metaphor of sedimentation suggests.

However, as I will show later, this view is problematic for two related reasons: first, the difference between the existential, historiographical, and geological levels is not simply one of scale or velocity; rather, the different times are *dislocated*, which means that the temporal space shared by these temporalities is not continuous. Second, the understanding of the different timelines as sedimented parts of the same "cone" suggests that the distinct times are contemporaneous, in the sense that they belong within one temporal space – a common present, if you will. The problem with a shared present, I will argue, is that it depoliticises theories of history committed to such a view. We should therefore propose an alternative model of historical time, capable of accounting for both the dislocation and the non-contemporaneity of temporal space.

Taubes observes that memory offers a possibility of dislocation between subjective time and worldly temporality. For Taubes, memory has a capacity to resist worldly time by its capacity to prevent events from disappearing:

> Memory uncouples an event from the stream of time. An event can be released from the time element in this way because it is set fast and does not disappear in the course of time . . . time is conquered by memory. Because memory stands *outside* time, it can be aware of its transitory nature.
>
> *(2009, 14)*

Furthermore, a similar non-identity can be found in the relation between historiographical and geological times. Recently, Dipesh Chakrabarty has drawn a distinction between "global" and "planetary" temporalities – while the former refers to human history, the latter involves "vast processes of unhuman dimensions":

> The global, as I have said, refers to matters that happen within human horizons of time, the multiple horizons of existential, intergenerational, and historical time, though the processes might involve planetary scales of space. Planetary processes, including the ones that humans have interfered with, operate on various time tables, some compatible with human times, others vastly larger than what is involved in human calculation. Thus air and surface water have "short recycling times," as do many metals, but soils and ground water take "thousands of years" to replenish themselves. . . . The two modes of thinking represent two different kinds of knowledge and, for humans, two different ways of comporting themselves to the world within which they find themselves. The global with humans at its center is ultimately all about forms and values. . . . But the planetary as such . . . cannot be grasped by recourse to any ideal form. There is no ideal form for the earth as a planet or of its history or for the history of any other planet.
>
> *(Chakrabarty 2019, 24–25)*

As Chakrabarty suggests, despite the intersection between the global (existential and historiographical) and the planetary (or geological) timelines, there is a more fundamental difference between these respective temporalities, namely the generation of two types of knowledges and comportments to the world. The temporal space between historiographical and cosmological temporality can be represented as dislocated according to the divergent epistemic and axiological effects produced by the respective macrocosmic timelines. If we now combine this insight with our analysis of memory – that is an experience proper to the subjective microcosm or existential time, which is (at least partly) exterior to both historiographical and geological temporalities, we can conclude that the relationship between the three times is discontinuous.[9]

One may argue that the dislocation of the three timelines does not prevent them from being *contemporaneous*. The gradually expanding circles of a cone may not perfectly overlap, but they still express a common present, running vertically through all the horizontal levels. The existential time would then be located within *the same moment* as parts of the historiographical and cosmological timelines, and it is this universal present which would account for the interconnectedness between the three temporalities.

In *Reading Capital*, Althusser provides reasons as to why temporal contemporaneity should be rejected. Althusser argues that philosophies in which "all the elements of the whole always coexist in one and the same time, one and the same present, and are therefore contemporaneous with one another in the present" (Althusser et al. 2015, 241) make it impossible to gain any knowledge of the future; this, in turn, makes such philosophies politically redundant:

> [T]he ontological category of the present prevents any anticipation of historical time, any conscious anticipation of the future development of the concept, any *knowledge* of the *future*. . . . The fact that there is no knowing the future prevents there being any science of politics, any knowing that deals with the future effects of present phenomena.
>
> *(Althusser et al. 2015, 242)*

The moment we situate all temporal levels within a contemporaneous temporal space constituted by a universal moment, we, in effect, assert that "nothing can run ahead of its time" – the limit of the present would be the limit of all three temporalities. This, in turn, makes it impossible to devise any reliable plans for political action: to anticipate possible future developments and, with them, opportunities for change, at least some temporalities would have to reach *outside* of the all-encompassing present, offering access to future alterations (or lack thereof), which, in turn, would illuminate the direction of other timelines, informing political action – which the levels, as contemporaneous, cannot do. Thus, if the knowledge of the future is in principle unavailable, we can only "divine it as a presentiment" (Althusser et al. 2015, 242), instead of attempting to prepare for, and shape what is coming. The belief in a universal present, therefore, depoliticises apocalypticism.

It is Althusser who offers an alternative model of historical time which, I believe, can be useful for conceptualising the relationship between the levels of apocalyptic history. Importantly for our purposes, such a model of time both politicises apocalypticism and allows us to address P2 – the second decolonial objection, which critiques eco-eschatology for being unable to account for the differentiation of the climate crises.

Althusser asserts the dislocation and the concomitant relative autonomy of multiple timelines.[10] However, he also recognises the interconnectedness of divergent temporalities:

> The fact that each of these times and each of these is histories is *relatively autonomous* does not make them so many domains which are *independent* of the whole: the specificity of each of these times and of each of these histories . . . is based on a certain type of *dependence* with respect to the whole.
>
> *(Althusser et al. 2015, 247)*

For Althusser, therefore, the temporal levels create a particular historical "conjuncture" in which they coexist in a relatively independent way and in which they are connected by relations of backwardness, forwardness, survival, and unevenness. The timelines, therefore, constitute the *non-contemporaneous* temporal space – "the whole" whose mechanism, in turn, determines the divergent timelines (Althusser et al. 2015, 254).[11] Thus, on the Althusserian model the relations of backwardness, forwardness, survival, or unevenness of different temporal levels are determined not by the base time or the universal present but by the relatively autonomous positions they assume *vis-à-vis* each other in a given historical totality (Althusser et al. 2015, 254). As Vittorio Morfino puts it, commenting on relevant passages of *Reading Capital*, time "is the articulation of a plurality of durations and at the same time the guarantee of the impossibility of hypostatisation of one rhythm in relation to the others. . . . It is, in a certain sense, the guarantee that time has no secret, that its fundamental structure is that of *non-contemporaneity*" (Morfino 2014, 15).

What are the consequences of representing history as dislocated and non-contemporaneous? First, we should not expect to find the element constitutive of one level in other levels: "The present of one level is, so to speak, the absence of another, and this co-existence of a 'presence' and absences is simply the effect of the structure of the whole in its articulated decentricity" (Althusser et al. 2015, 252). A similar point is made by Siegfried Kracauer in his discussions of the non-identity between macro- and micro-histories. Kracauer uses a cinematic analogy to demonstrate how the diverse effects of close-ups and long shots in film are mirrored by historical "close-ups," "apt to suggest possibilities and vistas not conveyed by the identical event in high magnitude history"[12] (1995, 126).

Furthermore, it is possible that the different levels run ahead or behind each other. If we take the existential and the historiographical timelines as an example, we can observe that an individual can be "ahead of their time" or "behind their times" and that historical events can be missed or anticipated by individuals (think here of a partisan in a forest who continues to fight because the news of the peace treaty hasn't reached them yet; or of a scientific discovery which cannot be integrated into current theoretical paradigms).

In an eschatological context, the Althusserian model of history would suggest that an analysis of ends of the worlds belonging to different levels would yield heterogeneous findings (analogous to the difference between macro- and micro-histories) and that the ends of the worlds would happen at different points in each timeline. However, the differentiated apocalypses would nonetheless belong together within an apocalyptic historical conjuncture, which would determine their positions in relation to each other.

The rather abstract characterisation of the temporal levels as constitutive of a dislocated and non-contemporaneous historical conjuncture has concrete consequences for politics – illustrated by Althusser with an example of the Russian Revolution. In "Contradiction and Overdetermination," Althusser suggests that the October Revolution was possible because "Russia was, *simultaneously at least a century behind the imperialist world, and at the peak of its development"* (1969, 97). Lenin's success, therefore, consisted of grasping and exploiting the relationships of temporal forwardness, backwardness, survival, and unevenness of multiple heterogenous timelines articulated in a decentred historical totality,[13] as "the *objective conditions* of a Russian revolution" to "forge its *subjective conditions*, the means of a decisive assault on this weak link in the imperialist chain" (1969, 98).

The political potential of the dislocated and non-contemporaneous model of time, therefore, consists of representing a complex historical totality in a differentiated way, paying attention to the relationships of backwardness, forwardness, survival, and unevenness between particular timelines. Political strategists, in turn, can exploit these interrelations in order to devise effective interventions aiming at transforming a given conjuncture. Furthermore, the thesis regarding the relative autonomy of timelines can also help us to both make sense of and address the survival of pre-transformational elements after the general mutation of the conjuncture (Althusser 1969, 115–116).

I believe that the account of historical time I sketched with the help of Althusser is able to address P2, that is the problem of covering over and depoliticising the differentiation of the climate crises. The dislocated and non-contemporaneous history (a) replaces a single eschatological timeline with multiple apocalyptic temporalities articulated in terms of multiple levels: existential, historiographical, and geological (and cosmological); (b) by resisting subordinating heterogenous timelines to a universal present or a base time it avoids constructing single extinction narratives – which makes it well-equipped to capture the unequal temporal differentiations of the climate crises; (c) it can inform environmental politics by representing apocalyptic times and narratives as a historical totality in a manner conducive to radical transformations.

Strategy and tactics

A critic may point out that the understanding of history as a spiral (which allowed us to address P1) and as a decentred totality (sketched in my response to P2) offers two seemingly incompatible shapes of historical time, which present history either as a *diachronic process*, where the intersection between past, present, and future generates the cyclical yet forward movement of time, or as a *synchronic structure* (albeit capable of breaks and mutations), where all elements constitute a historical whole. If this is the case, not only is eco-eschatology unable to offer a unified model of history capable of addressing both P1 and P2, but its solutions to P1 and P2 reveal it to be internally contradictory – since they demonstrate that eco-eschatology is committed to two irreconcilable shapes of historical time.

This objection can be answered by pointing out that (1) the synchronic totality consists of diachronic processes and that (2) the concept of a conjuncture can be a tool for conceptualising the relationship between spiral developments taking place in individual timelines constitutive of a historical totality. In other words, eco-eschatology has two *complementary* ways of thinking about historical time at its disposal: while the first one focuses on the spiral histories found in existential, historiographical, geological, and cosmological levels; the second one captures the complex coexistence of the spiral timelines in a given historical whole.

It should be noted that sharing of a spiral shape by individual timelines does not undermine their dislocation, since the times remain qualitatively distinct: the spiral relationship between past, present, and future has a different quality in the case of one's psychological life than in the case of a spiral recurrence of historical or geological events. For instance, the repetition of a traumatic memory – a personal end of a world – differs qualitatively from a revolutionary repetition or a recurring of natural disasters.

Importantly, the twofold approach to eschatological history maps onto the distinction between *tactics* and *strategy*, which makes it useful for eco-politics. The focus on individual spiral histories enables us to come up with tactical political solutions shaping the development of a given timeline. By contrast, the meta-perspective representing the historical conjuncture allows us to devise an overall strategy capable of transforming a given totality. Furthermore, the decision on the usefulness of particular tactics (or on the usefulness of the resistance to their application) can be made only from a strategic point of view, grounded in the synchronic meta-perspective. In addition, the grasp of a conjuncture may reveal a need for new tactics, which would respond to political demands unseen from within particular timelines.

More specifically, by representing multiple apocalypses as part of a dislocated and non-contemporaneous totality, we can differentiate between different forms of reproduction and non-reproduction proper to each timeline. Consequently, we can "play" apocalypses against each other, stopping some and precipitating other ends of the world, responding to their qualitatively distinct nature. In short, eco-apocalyptic politics grounded in a dislocated and non-contemporaneous view of history can account for the "heterogeneous temporality of the Anthropocene" (Rothe 2020, 163); it can also generate tactics and strategies, which, while aiming at non-reproduction – can lead to a change of the historical whole and an emergence of a new *status*.

Notes

1 While the *content* of Joachim's and Taubes's respective eschatologies can be accused of Eurocentrism, as I show later they develop a *formal* account of history compatible at least with some demands of decoloniality.
2 See McGinn (1979a, 133–134).
3 The image of Table XI from *Liber Figurarum* representing Joachim's vision of history can be found online.

4 Alternatively, Marjorie Reeves suggests that, perhaps, it was the "annually return-ing order of the Church's liturgy" which inspired Joachim usage of cyclical ele-ments in his account of history (1980, 270).

5 "The dialectic of Paul's history of salvation is both *quantitative in terms of world history* (Hegel) and *qualitative in existential terms* (Kierkegaard)" (Taubes 2009, 63). In another work, Taubes explains the parallels between the texts of St Paul and Walter Benjamin by suggesting a concordance of their experiences (2004, 72–74).

6 "Apocalypticism and Gnosis inaugurate a new form of thinking. . . . The logic of the dialectic, whether in apocalypticism and Gnosis or in the works of Hegel, is not circular but spiral. The 'bending backward' characteristic of the dialectic does not progress back to the thesis in a circular manner but broadens out into spiral toward the synthesis. . . . Dialectic logic is a logic of history, giving rise to the eschatological interpretation of the world" (Taubes 2009, 35). For a more detailed discussion of the relationship between apocalypticism, Gnosis, Hegel, and Taubes, see Agata Bielik-Robson, *Jewish Cryptotheologies of Late Modernity* (2014), esp. chapter 5.

7 "A new epoch is beginning, which introduces a new aeon that is post-Christian in a more profound sense than that of the calendar. This epoch, in which the threshold of Western history is crossed, regards itself primarily as the no-longer of the past and the not-yet of what is to come. . . . For the coming age is not served by demonizing or giving new life to what-has-been, but by remaining steadfast in the no-longer and the not-yet, in the nothingness of the night, and thus remaining open to the first sings of the coming day" (Taubes 2009, 193).

8 Of course, the concordance between Joachim and Taubes is far from acciden-tal – *Occidental Eschatology* contains a relatively long section devoted to Joachim, which demonstrates Taubes's in-depth engagement with the thought of the medi-eval monk.

9 It would be interesting to read the notion of dislocation into Joachim's texts – such an interpretation would then split the history of the three *status* into heterogenous histories taking place on the levels of characters, events, and the Godhead. This, in turn, would complicate the concept of a *single status*.

10 "Each of these different 'levels' does not have the same type of historical existence. On the contrary, we have to assign to each level a *peculiar time*, relatively autono-mous and hence relatively independent, even in its dependence, of the 'times' of the other levels" (Althusser et al. 2015, 246–247).

11 Interestingly, as Lynch points out, the notion of "non-contemporaneity" is impor-tant for Ernst Bloch's apocalypticism (Lynch 2019, 111). This suggests that an analysis of time can offer an interesting point of convergence between Althusse-rian and Blochain projects.

12 In a similar fashion, Carlo Ginzburg notes: "A battle, strictly speaking, is invis-ible, as we have been reminded (and not only thanks to military censorship) by the images televised during the Gulf War. Only an abstract diagram or a visionary imagination . . . can convey a global image of it. It seems proper to extend this conclusion to any event and with greater reason to whatever historical process. A close-up look permits us to grasp what eludes a comprehensive viewing, and vice versa" (1993, 26).

13 Among the examples constitutive of the historical conjuncture in 1918 listed by Althusser, we find the historical contradictions between feudalism and "large-scale capitalist and imperialist exploitation"; the industrial cities and "the medieval state of the country side"; the class struggle within the ruling classes, and "the '*advanced*' character of the Russian revolutionary élite . . . *far ahead of any West-ern 'socialist' party in consciousness and organisation*" (1969, 96).

Bibliography

Althusser, L. (1969) *For Marx*. Trans. B. Brewster. Harmondsworth: Penguin Books

Althusser, L., et al. (2015) *Reading Capital: The Complete Edition*. Trans. B. Brewster & D. Fernbach. London: Verso

Bielik-Robson, A. (2014) *Jewish Cryptotheologies of Late Modernity: Philosophical Marranos*. London: Routledge

Brannen, P. (2017) *The Ends of the World: Volcanic Apocalypses, Lethal Oceans and Our Quest to Understand Earth's Past Mass Extinctions*. New York: Simon and Shuster

Brassier, R. (2007) *Nihil Unbound: Enlightenment and Extinction*. London: Palgrave Macmillan

Chakrabarty, D. (2019) "The Planet: An Emergent Humanist Category," *Critical Inquiry* 46, pp. 1–31

Danowski, D., Viveiros de Castro, E. (2016) *The Ends of the World*. Trans. R. Nunes. Cambridge: Polity Press

Derrida, J. (1994) *Spectres of Marx: The State of the Debt, the Work of Mourning and the New International*. Trans. P. Kamuf. London: Routledge

Dupuy, J.-P. (2015) *A Short Treatise on the Metaphysics of Tsunamis*. Trans. M.B. DeBevoise. East Lansing: Michigan State University Press

Dupuy, J.-P. (2022) *How to Think About Catastrophe: Toward a Theory of Enlightened Doomsaying*. Trans. M.B. DeBevoise & M.R. Anspach. East Lansing: Michigan State University Press

Eliade, M. (1959) *Cosmos and History: The Myth of the Eternal Return*. Trans. W.R. Trask. New York: Harper & Brothers

Freud, S. (2003) *The Uncanny*. Trans. D. McLintock. London: Penguin Books

Ginzburg, C. (1993) "Microhistory: Two or Three Things That I Know about It," Trans. J. Tedaschi & A.C. Tedaschi. *Critical Inquiry* 20, pp. 10–35

Koselleck, R. (2018) *Sediments of Time: On Possible Histories*. Trans. S. Franzel & S.-L. Hoffmann. Stanford: Stanford University Press

Kracauer, S. (1995) *History: The Last Things Before the Last*. Princeton: Markus Wiener Publishers

Marx, K. (1972) *The Eighteenth Brumaire of Louis Bonaparte*. Moscow: Progress Publishers

McGinn, B. (1979a) *Visions of the End: Apocalyptic Traditions in the Middle Ages*. New York: Columbia University Press

McGinn, B. (1979b) *Apocalyptic Spirituality*. New York: Paulist Press

Mignolo, W.D. (2021) *The Politics of Decolonial Investigations*. London: Duke University Press

Morfino, V. (2014) *Plural Temporality: Transindividuality and the Aleatory Between Spinoza and Althusser*. Leiden: Brill

Ranawana, A., Trafford, T. (2019) "Imperialist Environmentalism and Decolonial Struggle," *Discover Society*. Available online: https://archive.discoversociety. org/2019/08/07/imperialist-environmentalism-and-decolonial-struggle/ [Accessed 23/2/2022]

Reeves, M. (1980) "The Originality and Influence of Joachim of Fiore," *Traditio* 36, pp. 269–316

Reeves, M. (1999) *Joachim of Fiore and the Prophetic Future*. Guildford: Sutton Publishing

Regalado Fernandez, O.R. (2023) "On the Apocalyptic Theme in Modern Scientific Discourse," in *The Environmental Apocalypse: Interdisciplinary Reflections on the Climate Crisis*, J. Kowalewski (Ed.). Abingdon: Routledge

Rothe, D. (2020) "Governing the End Times? Planet Politics and the Secular Eschatology of the Anthropocene," *Millennium: Journal of International Studies* 48, pp. 143–164

Tanaka, M. (2014) *Apocalypse in Contemporary Japanese Science Fiction*. London: Palgrave Macmillan

Taubes, J. (2004) *The Political Theology of Paul*. Trans. D. Hollander. Stanford: Stanford University Press

Taubes, J. (2009) *Occidental Eschatology*. Trans. D. Ratmoko. Stanford: Stanford University Press

Wolin, R. (2023) *Heidegger in Ruins: Between Philosophy and Ideology*. New Haven: Yale University Press

Zylinska, J. (2018) *The End of Man: A Feminist Counterapocalypse*. Minneapolis: University of Minnesota Press

5

A REQUIEM FOR A WORLD BUILT ON SAND

Landscapes and the ambivalence of ruins

The concept of "the world" has been under attack. Thomas Lynch notes two main sets of criticisms waged against it: the first one is philosophical, showing that "the world" is a theoretically incoherent notion; the second one is political, equating "the world" to "an imperialist attempt to impose a unitary imaginary on a planet that is home to many human and other-than-human worlds" (Lynch 2024, 22). The affinity between these two criticisms can be established by experience: the incoherence of the world implies that it *can't* be experienced; if we believe that every day we encounter the world, it is only because its concept has been imposed on us by imperialist power structures.

However, giving up on the idea of the world – whether on philosophical or political grounds – is a slippery slope for any apocalyptic project; the absence of the world could make apocalypticism redundant, since, as Lynch points out, discussions "of the end of the world presume that there is something called the world that could end" (2024, 22). Operative in this worry is a conviction that apocalypse presupposes the world as its condition – a belief grounded in an intuitive metaphysical dictum that for X to end, there must first be an X. Now, if apocalypse is *secondary* to the world, both ontologically and conceptually, then the relevance and usefulness of apocalypticism depends on defending the notion of the world. In short, no end of the world without the world!

But what if the commonsense metaphysics misleads us, and, at least sometimes, the end not only *coexists with* but also *precedes* X? What if, contrary to our intuitions, the world *presupposes* apocalypses? It is climate apocalypse which today enables us to posit such a hypothesis: to say that the catastrophic effects of global warming produce new (albeit worse) environmental

DOI: 10.4324/9781003348511-6

conditions of existence, and that this new world should be understood *starting from* climate apocalypse, is far from absurd.

One way to conceptualise the reversal of the relationship between the world and its end is by thinking of these terms as corresponding to two types of thresholds. Cezary Rudnicki distinguishes between a threshold as a *border* which both separates and unites two distinct contexts (e.g. the border between two rooms in a house) and a threshold as a specific *degree of intensity* internal to a system, whose quantitative increase can effect a qualitative transformation (e.g. heating up of a piece of metal turns the solid object into a liquid) (2024, 143–144). What I would like to suggest is, first, that *the world* is a border which separates and unites distinct apocalypses, that is it is a threshold in between ends, capable of differentiating and connecting catastrophes. For instance, today's state of the world could be described as the border between climate, economic, and nuclear ends of the world. Second, *apocalypses* stand for a specific degree of intensity internal to the world; they are thresholds with a power to cause qualitative shifts through quantitative increments. The multiple apocalypses which populate the past and the present – including climate, economic, and nuclear ends of the world – structure the world *from within*, delimiting and generating its possible states.

While on the first understanding, the world and its end appear mutually constitutive, at least conceptually (there is no threshold without two states, but, conversely, there are no two states without a threshold); the second reading places an emphasis on the ontological priority of apocalypse: the world can function as a border between ends only because these apocalypses are *internal* to the world, structuring the latter from within as an intensive principle explaining both the maintenance of and a change in the state of the world. In both cases, however, *apocalypses cease to be secondary to the world*. In short, no world without the end of the world!

In this chapter, to demonstrate the possibility of life *after* the end of the world, I will examine the examples of subjective world-breakdowns, as well as urban and forest landscapes structured by multiple apocalypses. The analysis of landscapes, in turn, will enable me to consider the constitution of space by temporal processes populated with apocalypses and the special status of the economy in the apocalyptic production of the world. I will conclude this chapter by arguing for the normative ambivalence of apocalypses, which puts into question any attempts to situate the ends of the world in stories of redemption. Throughout the chapter, I will also address the two criticisms of *the world*, arguing, first, that the world is experienceable, and that its concept is philosophically coherent; and second, that the opposition between one and many worlds, operative in the political critique, is a false dichotomy – the imperialist world is produced *in* and *as* many worlds.

A phenomenology of (losing) the world

What is the world? It could refer, among other things, to our immediate environment, or to planet earth, or – in philosophical discourses – to a totality of beings. Importantly, asking *what* the world is already frames the parameters for a possible answer, since it presupposes that the world is an *object*. Admittedly, it is a unique object, distinguishable from things such as tables or tress, but it is an object nonetheless: an environment, a planet, a totality etc.

However, as Anthony Steinbock argues, we can also approach the question of the world from another angle – rather than asking *what* it is, we should ask "*how* as lifeworld it is always already at play" (1995, 103). Such an approach would require explicating the world through a phenomenological enquiry into *how* the world structures our reality.

Steinbock proposes to articulate the phenomenology of the world in terms of its two modalities: "world-horizon" and "earth-ground." Drawing on Husserl, Steinbock characterises the world-horizon as *the transcendental condition of the appearance of objects*. More specifically, world-horizon reveals by means of "*referential implications*": "the way in which one thing points toward another, or the way in which sense is implied" (1995, 107). At any point, we are surrounded by objects – for example buses, cars, roads, pavements, shops, restaurants, parks, and even the weather – which refer to each other, constituting a meaningful environment: if it isn't raining, I will choose to walk to my destination; when doing so, I will stick to the pavement and avoid walking on a busy road; when I get tired I would sit in the park, either on the grass or on a bench in the park; if it starts raining, I will hide in the pub. As this everyday example indicates, the world-horizon – by allowing things to make sense as part of a meaningful whole – makes possible practical engagement with the world around us: as Steinbock puts it, the world-horizon implies appropriate "styles" of engaging with objects (Steinbock 1995, 109). For example, in the scenario above, the meaning of referential implication, and consequently the way in which I navigate this environment, would be different if I am on a lunch break or if I'm enjoying a day off, if only in terms of having to keep time in the former case. World-horizon, therefore, both opens up possibilities (by meaningfully connecting objects) and delimits possibilities (by implying appropriate praxis) (Steinbock 1995, 109).

Importantly, world-horizon, while allowing for the appearance of things, is not itself an object. The interconnectedness of things responsible for the making sense of our world "withdraws *in favor of the object being given*," that is, it recedes behind the very things whose appearance it makes possible. In consequence, "the horizon escapes substantial thematization" (1995, 107).

The second transcendental condition of our experience of the world is the earth-ground. If the world-horizon establishes a referential network of meaning, the earth-ground "structures our experience of space and move-ment" (Steinbock 1995, 113). The work of the earth-ground was already implicit in the previous account of going for a walk: the spatial possibilities afforded to me by my environment and my body, for example, the fact that I could *walk* on a pavement or *sit* on the grass and that I could be *exposed to bad weather* in a park or be *hidden from rain* in a pub, are all part of the reve-latory power of the earth-ground. In the most general sense, the earth-ground names the way in which meaningful connections between things unravel rela-tive to specific possibilities of rest and movement in space – whether that's the movement of my body, of cars, or of rain drops. On Steinbock's re-reading of Husserlian phenomenology, therefore, human bodies, but also the bodies of animals, plants, or bacteria, insofar as they predisposed to move and rest relative to earth-ground would "gain their *constitutive* sense *from the earth*" (Steinbock 1995, 119).

> The earth belongs not *merely* in a biological, psychological, or ecological register for Husserl, but in a *transcendental* register, since it is a condition for the possibility of sense-giving; it functions as the pregiven ground, as that from which things ultimately take on sense; it gives sense even to my lived-corporeality, my lived-subjectivity that is itself and in its own way sense-giving.
>
> *(Steinbock 1995, 120)*

Note that on this understanding, the earth-ground is primarily a condition of making sense; the notion of the earth as a planet, a round-ish object which is part of the solar system, is only *secondary* to earth-ground as the primordial ground of meaning. In fact, we can make sense of earth as a planet because of the combined revelatory power of the world-horizon and earth-ground.

Now, the world can appear *familiar* or *strange*. To stick to our example of a walk, if I stroll in an area I know well, the implied style of engage-ment with objects makes sense to me. By contrast, if I visit a new area, say, in a foreign country, it will take me a while to work out how exactly to engage with the possibilities afforded by the environment – can I sit on this piece of grass or would this be frowned upon?; do people queue before they get on a bus?; what are the signs communicating in the local language? etc. Steinbock proposes to call contexts which are normatively familiar *homeworlds*. They are constituted through a process of normative appropriation – concepts, values, norms, rituals, and traditions articulate the normative structures of a given context. It is this work of appropria-tion, that is the way we take up the normativity of our environment, which enables us to *make ourselves at home in the world*. (As I will argue in the next chapter, in addition to normative practices, homeworlds are also con-stituted by material appropriation.)

The existence of homeworlds suggests, first, that the normative structures of the homeworld are never fixed: insofar as the homeworld presupposes normative (and material) appropriation, it "is *modified in the repeating*, that is, it is generated" (1995, 199). Second, every homeworld implies an *alien-world*: a context which would be normatively opaque and materially inaccessible. In fact, as Steinbock points out, homeworlds and alienworlds are "*co-relative* and *co-constitutive*" – the constitution of one is the delimitation of the other (1995, 179).

The phenomenological account of the world allows us to begin addressing the first, philosophical critique of the world: the world is a coherent notion precisely because it makes possible experience through its two modalities of the earth-ground and world-horizon, whose specific appropriation generates homeworlds.

Importantly for our purposes, Steinbock acknowledges the possibility of a homeworld's breakdown: "a home is no longer a home when the sense of the homeworld is no longer appropriated, when it is no longer identifiable as *our* world" (1995, 235). My ability to constitute a homeworld can be disrupted by, for example:

> the death of a child, which can change the entire horizon of interests and thus the mode of life of the parents; the death of a spouse as a co-bearer of a communal life tasks; the dismissal of co-workers with whom we have committed our communal life goals; a war which calls into question the entire future of a homepeople or a home nation; natural disasters such as floods or earthquakes, which upset our entire normal or communal homeworld; the break of a tradition as in forgetting, and finally the intervention of others in our lives, which alter the unfolding of our future experience.
>
> *(Steinbock 1995, 241)*

For Steinbock, however, the cases of world-breakdown are only *secondary* modifications of the typical experience of the world. A more fundamental disintegration of the world has recently been argued for by Timothy Morton.

According to Morton, we experience the world as an implicit background of objects: "a container in which objectified things float or stand" (2013, 99). This worldly background, however, is only "an objectification of a hyper-object: the biosphere, climate, evolution, capitalism" (2013, 100). Today, global warming has shifted the hyperobject *weather* from its background position to the foreground, fundamentally disrupting our experience of the world in two ways. First, the world becomes *strange* because it no longer possesses the familiar structure of foreground and background. Second, the breakdown of the background–foreground structure which, for Morton constitutes our experience of the world, in effect marks *the end of the world*: the latter "has evaporated. Or rather, we are realizing that we never had it in the first place" (2013, 101). Without the foreground and background structure, Morton tells us, the world can't be experienced; all that we *can* experience are the movements of hyperobjects, which change positions in relation to each other.

I believe that Morton's account of the disappearance of the world relies on an incorrect phenomenology – Steinbock is clear that the modalities of the world are "not to be mistaken for the background or a foreground" (Steinbock 1995, 108). Neither the earth-ground nor the world-horizon is an object, not even Morton's *hyperobjects*; rather, they are *conditions of experience*, that is they make possible both the meaningful appearance of (hyper) objects and the changing spatial differentiation between things "close by" and "far off." Nonetheless, I think that Morton intuits something extremely important: our *typical* experiences are constitutively structured by apocalypses or disappearance of worlds; consequently, even the most mundane experiences may presuppose the end of the world. However, the reason for the apocalypticism of everyday life is neither a hyperobject nor the disintegration of the foreground–background structure but a regular collapse of the transcendental foundations of my homeworld: the earth-ground and the world-horizon. In the words of the Gospels, making ourselves feel at home in the world is like building an abode on sand rather than rock: "the rain fell, and the floods came, and the winds blew and beat against that house, and it fell, and great was the fall of it" (Matt 7:27).

World-breakdowns: death of the other, cruelty, colonialism

What does the *experience* of world-breakdown look like? As I will demonstrate, apocalyptic experiences follow a two-stage pattern: first, an event or a process causes the disintegration of the transcendental conditions of experience (i.e. the world-horizon and earth-ground); second, the subjects attempt to recuperate the world by means of normative and material appropriation of the breakdown. Here I will examine three experiences of world-breakdown, caused, respectively, by the death of the other, cruelty, and colonialism. These three experiences will also mark a transition from events and processes which can be integrated in the post-apocalyptic world (and which we may call *minor* apocalypses), to events and processes which refuse such a reintegration (which could be called *major* apocalypses). In the next section, I will argue that the subjective experiences of world-breakdown take place in particular spatiotemporal contexts, which themselves are structured by apocalypses.

Jacques Derrida famously compares the death of the other to the end of the world: death "marks each time . . . the absolute end of the one and only world." The lonely survivor who outlives a fellow being is "in some fashion beyond or before the world itself" (2005, 140). We can articulate further Derrida's intuition by identifying two related aspects of the world-collapse caused by the other's death.

First, a death of another being – especially one close to us – causes a breakdown of the referential network of meaning. Others are a significant point of reference for my sense-making, that is family members, friends, and even pets are implied in how objects point to each other (e.g. a park is a place where I walk my dog or play with my child). Consequently, the sudden absence of the other disorients this horizon of meaning – objects *try to* but are unable to refer in a familiar way; the disappearance of another being pierces a hole in the referential network, and the world, at least in part, ceases to make sense.

Second, the other – especially a stranger – is a source of an alienworld which, as Steinbock argues, marks the boundaries of my homeworld, co-constituting it. The breakdown of the alienworld, which the death of the other enacts, therefore, would result in the dissolution of the normative boundaries of my homeworld. In other words, if the other's world marks the limits which envelope my world – manifested in my experiences of familiarity and alienness – then the death of the other would dissolve these limits, confusing familiar styles of engaging with the world. In literature and film, the strangeness which follows the disappearance of the alienworld is represented by the trope of a single survivor on a lonely island – the hero is lost and confused, not just geographically and psychologically but also phenomenologically, since their homeworld has become boundless, and the categories of familiarity and alienness no longer apply in the same way. A world without others is in a significant sense the end of the world for me.

Of course, the survivor is not lost forever, nor is the subject who outlives a fellow being. The survivor, therefore, continues living, albeit in the "world after the end of the world" (Derrida 2005, 140). As Derrida points out, the death of the other becomes normatively appropriated: we experience the demand to *take on* and *carry* the other's world (Derrida 2005, 140). We can read this ethical injunction as attesting to a possibility of reconstituting our world in two ways: first, we must preserve the other as a reference in our network of meaning, albeit in a transformed way (e.g. by visiting their grave or remembering them on important dates); second, we must reintegrate the remnants of their alienworld into our homeworld, for instance, by assuming the norms dear to the dead other. As Timothy Secret aptly puts it:

It is perhaps not an accident that the visual archetype of the mourner is a figure bent double like Atlas bearing the world on their back, since this is precisely how we are left in the wake of the other's death – carrying their entire world now they are no longer there to carry it for themselves.

(2015, 199)

A more extreme case of world-breakdown can be observed in the experience of cruelty. As Lynne S. Arnault observes "cruelty, like other serious harms, involves the unmaking or near unmaking of a person's sense of self and world" (Arnault 2003, 165). Since cruelty, in addition to causing material harm, also violates our norms (Arnault 2003, 168), we can understand the experience phenomenologically as an *interruption* of the material and normative processes of appropriation, responsible for the generation of my homeworld. When I experience cruelty, the norms which enable me to navigate my world are violently put into question; furthermore, the physical and psychological effects of cruelty often impact my experience of spatial possibilities of rest and movement – there are places I no longer feel able to walk through or stay in. In other words, cruelty affects both my horizon of meaning and my experience of the earth-ground, disintegrating my homeworld.

Sometimes, when attempting to reconstitute their world, the survivor will attempt to normatively appropriate the experience of cruelty by integrating it into a *narrative of redemption* by "believing that the suffering caused by cruelty must have some good or transcendent purpose, we implicitly express our resolve to remain faithful to our normative prescriptions" (Arnault 2003, 173). However, as Arnault points out, not only are redemption narratives politically suspect (I will return to this point in the latter part of the chapter), they may also be futile. The end of the world experienced by the subject may forever resist appropriation, leaving the constitution of the homeworld forever unfinished.

> The self that arises from the ashes may be a lost or shattered self, sometimes even one whose will and character have become infected by cruelty. Cruelty can be "senseless," not in the (mistaken) sense of being motiveless, but in the sense of resisting the victim's attempts to make it part of a life story that has sense and a coherent unity.
>
> *(Arnault 2003, 166)*

The third apocalyptic experience I would like to consider is one produced by colonialism. As we have seen in previous chapters, the "discovery" of the so-called New World was, in fact, the end of the world for the indigenous population of the Americas; the Amerindian world was disintegrated through colonial subjugation, enslavement, epidemics, and genocide (Danowski & Viveiros de Castro 2016, 107). In addition, the colonial economy transformed indigenous landscapes through the extraction of resources, implementation of new agrarian systems, building of settlements, and establishment of trade outposts, with significant consequences for local eco-systems (Lightfoot et al. 2013). In consequence, "the surviving Indians . . . found themselves as *humans without world*: castaways, refugees, precarious lodgers in a world in which they no longer belonged, because it could not belong to them" (Danowski & Viveiros de Castro 2016, 105–106).

Insofar as colonialism involves genocide and violent subjugation, it would intensify the processes we identified in our analysis of death of the other and cruelty – albeit with important differences. First, in contrast to the death of the other and cruelty undergone by an individual, the colonial apocalypse is a collective experience, extended over a centuries-long period of time. Second, if in the aftermath of the death of the other we integrate their world into our homeworld, in the case of colonialism the alienworld subsumes and disintegrates the indigenous world; consequently, rather than taking up and carrying the other's alienworld, the survivors "hold on to whatever little world is left to them" (Danowski & Viveiros de Castro 2016, 107). Third, while both cruelty and colonialism, by violently disorienting our referential networks of meanings, transform the spatial possibilities of rest and movement, in the case of colonialism the earth-ground is also modified by *environmental* alterations. In consequence, the indigenous population "carried on in *another world*, a world of others, their invaders and overlords" (2016, 106).

The extent of the world-breakdown effected by colonialism, compounded by its antagonistic character (attested by the fact that a homeworld is being destroyed by the alienworld), means that the world-building in the aftermath of colonial apocalypse can take the form of a struggle *against* the alienworld – as the example of the Zapatistas illustrate:

> [E]ven though they have gone through successive ends of the world, even though they have been reduced to a poor and oppressed peasantry, have had their territory broken up and handed over to different nation-states . . . the Maya continue to exist, their population grows, their world resists: diminished but defiant. And it is indeed the Maya who offer us today what may be the best example of a "successful" popular insurrection against the two-headed State-Market monster that oppresses the world's minorities. . . . We speak, of course, of the Zapatista uprising in Chiapas, that rare revolt that is a model of "sustainability" – political sustainability, also and above all. The Maya, who lived through their various ends of the world, show us today how it is possible to live after the end of the world.
> *(Danowski & Viveiros de Castro 2016, 107–108)*

The death of the other, cruelty, and colonialism enable us to think the world *starting from* its end. However, one may argue that none of the three examples of world-breakdown support my claim that apocalypses *precede* the world – in all three cases, the world pre-existed its end. To substantiate the thesis that apocalypses are prior to the world, I will argue that the spatiotemporal context in which the experiences of world-breakdown take place is itself structured by apocalypses. As I will show in the next section, it is the analysis of landscapes which makes legible the power of ends to ruin, generate, and stabilise spaces.

Landscapes or the ecologies of ends

Landscapes, according to W.J.T. Mitchell, should be conceived not as a noun but as a *verb* – a landscape is not only "an object to be seen or a text to be read, but . . . a process by which social and subjective identities are formed" (2002, 1). I would like to take up the characterisation of landscape as a process; more specifically, as I aim to demonstrate in this section, landscapes are situated on an intersection of existential, historiographical, and geologies times populated with apocalypses – timelines of processes responsible for the ruination, generation, and stabilisation of a given space. Admittedly, the apocalyptic constitution of a landscape is often not legible, since a "landscape effaces its own readability and naturalizes itself" (Mitchell 2002, 2). Nonetheless, as I will show here with the help of the examples of urban and forest spaces, every landscape possesses an *ecology of ends*: a specific composition of temporal apocalypses, which structure spatial possibilities of rest, movement, and meaning.

My first example is urban scenery of Szombierki – a district of Bytom, a city located in the Upper Silesia region of Poland, known historically for its coal mines and industrial plants.

There is a legend about Bytom: according to the story, two priests were murdered sometime in the 14th century by the city's inhabitants. This crime led the Bishop of Kraków, and then the Pope itself, to place a curse on the city – every time Bytom lifts itself up from disrepair, it shall fall into an even deeper ruin. True or not, the story offers a key for engaging with the city's landscape.

If you ever visit Szombierki, you may be able to find a spot from which the historical processes which intersect in this space become visible. Looking straight ahead, you will see the formidable Krystyna winding tower, a relic of a now-defunct coal mine (the tower, which resembles a hammer or a cross, before 1945 was known by its German name: *Förderturm Kaiser-Wilhelm-Schacht*). Behind you, you will find a housing estate built by the communist government in the 1960s and 1970s; while on your right, you will see a Western supermarket which appeared after the Polish transition to market economy. The area is overlooked by a church of St Margaret to your left, built on a hill of the same name – a place of origin of Bytom in the 11th century. The current building is a fourth reiteration of the church, which in the past was destroyed and reconstructed, including after a Hussite invasion. Next to the church is a cemetery.

Standing in this spot, you may initially think that the landscape is immobile – perhaps with the exception of the passing cars and people. Nothing could be further from the truth; the landscape embodies historical processes, which continue to shape the space. The Krystyna tower, and the abandoned post-industrial space which surrounds it (left after most buildings of the mine were disassembled, currently overgrown by modest amounts of grass and weeds) are the monumental reminders of the discovery and exploitation of coal and eventual closing of the mine: the *end* of the pre-industrial ecosystems and the subsequent *end* of the economic and social life animated by coal industry. The Polish name of the tower marks the transfer of Bytom from Germany to Poland, after the change of borders following the Potsdam conference. The tower, therefore, functions also as a memorial to the *end* of German admin-istration and the concomitant displacement of the German population after World War II. The sense of endings is compounded, on the one hand, by the juxtaposition of the communist housing estate and the Western supermar-ket, which together express the *end* of communism in Poland and the *end* of hopes associated with the transition to capitalism – whose failures are palpable everywhere in Bytom. On the other hand, the reassuring presence of the church of St Margaret and the cemetery attached to it are underwritten, respectively, by distant historical *ends* and the recent *ends* of individual lives. Note that the landscape has a direct impact on both the world-horizon and the earth-ground: ecologies of ends shape both the referential implications of objects – the fact that they meaningfully point to each other – and the spatial possibilities of rest and movement (who moves where and when).

If the Szombierki district illustrates a *condensation* of ends in one space, forest landscapes – as Anna Lowenhaupt Tsing argues – demonstrate the *dispersion* of connected ends across the planet, as an effect of economic processes.

Capitalism shapes local landscapes in two related ways: first, a particular area can be transformed to ensure a more effective extraction of resources and an increase in wealth. Second, when a particular resource can no longer be extracted, or when the opportunity to generate wealth is identified else-where, the area is abandoned and left in ruin.

> The timber has been cut; the oil has run out; the plantation soil no longer supports crops. The search for assets resumes elsewhere. Thus, simplifica-tion for alienation produces ruins, spaces of abandonment for asset pro-duction. Global landscapes today are strewn with this kind of ruin.
>
> *(Tsing 2015, 6)*

Tsing draws on the example of forest landscapes in Japan and Oregon to show how, across the 20th century, related economic processes led to the ruination of both spaces. These include the promotion of industrial forestry, the import of Oregon's wood to Japan, the subsequent access to cheaper tim-ber from other markets, the drop of domestic prices of wood in the US and

Japan, and the eventual abandonment of both the Oregonian and Japanese forests (2015, 207–211). The shared history of forest landscapes in Oregon and Japan and, specifically, their ruination after the withdrawal of capital demonstrates that capitalism has the power to link "the most geographically, biologically, and culturally disparate forests . . . in a chain of destruction" (2015, 212). Analogously to historical concordances (which, as I argued in the previous chapter, bypass chronological time by establishing meaningful relationship between distant temporal moments), the apocalyptic time of the economy can connect disparate landscapes, ignoring spatial distance. As Ting puts it in a conversation with fellow anthropologists: "any local bunch of grass that happens to be growing in a particular place is intimately linked to these long-distance movements of money, capital, property, rights, and law" (Latour et al. 2018, 594).

Importantly, Tsing recognises that the abandoned and ruined industrial forests create *ambivalent* conditions of life: for "some insects and parasites, ruined industrial forests proved a bonanza. For other species, the rationalization of the forest itself – before ruination – proved disastrous" (Tsing 2015, 211). To capture the "challenge of living in that ruin, ugly and impossible as it is" (2015, 213), Tsing introduces a distinction between *three natures*. The first nature designates ecological relations; the second nature refers to the transformation of the environment by capitalism; and the third nature stands for "what manages to live despite capitalism" (Tsing 2015, viii).

I believe that we should extend Tsing's concept of third nature to apply also to contexts which either predate or can't be reduced to capitalist apocalypses. While forests of Oregon and Japan demonstrate how life can thrive despite economic destruction, other landscapes – insofar as they are situated on the intersection of multiple apocalyptic timelines – preserve the memory of life in the aftermath of other ends of the world (think here of the church of St Margaret which "commemorates" the destruction accompanying the Hussite invasion). In fact, as the legend of Bytom's curse demonstrates, life in the midst of ruination was explicitly acknowledged centuries before the emergence of capitalism. The broadening of the concept of third nature would, in effect, amount to elevating the ambivalent possibility of living in apocalyptic ruins to a *trans-historical* possibility: on the broader understanding of the concept, the third nature would also be our first nature.

The trans-historical possibility of living in ruins, in turn, captures the relationships between the experiences of world-breakdown and landscapes understood as ecologies of ends. The two-stage experience of a breakdown and the reconstitution of a homeworld, examined in the previous section, doesn't take place in a vacuum; rather, it requires as its condition a spatiotemporal context: a landscape, itself constituted by multiple ends, in which a homeworld is disintegrated and reappropriated. The relation between experience and landscape is, therefore, twofold. On the one hand, ecologies of ends function as the always-already apocalyptic spatiotemporal contexts for new experiences of the end of the world; on the other hand, apocalyptic landscapes offer the possibility of living in ruins, which is echoed in the subjective desire to re-establish the homeworld after its breakdown, that is to continue living in the world after the end of the world.

This analysis enables us to further qualify our philosophical concept of the world. The latter can, indeed, be understood as the world-horizon and the earth-ground. However, these two modalities of the world are undermined, first, by experiences of world-breakdown (which attests to the possibility of disintegration of both the world-horizon and the earth-ground), and second, by historical apocalypses which structure spaces, altering referential networks and the spatial possibilities of rest and movement. This would suggest the following conclusion: the world is located in between two sets of ends of the world – those affecting experience and those shaping landscapes.

The becoming-space of time

Implicit in the previous section was a hypothesis that time (and specifically, apocalyptic time) can constitute space. In this section I will examine in more detail the ontological processes, which we can designate as "the becoming-space of time" (Derrida 1997, 68). As I will demonstrate, apocalyptic time possesses a triple power of creating, destroying, and stabilising spaces.

The priority of time over space has recently been expressed by Pope Francis. In *Evangelii Gaudium*, Francis writes:

> Broadly speaking, "time" has to do with fullness as an expression of the horizon which constantly opens before us, while each individual moment has to do with limitation as an expression of enclosure. . . . One of the faults which we occasionally observe in sociopolitical activity is that spaces and power are preferred to time and processes. Giving priority to space means madly attempting to keep everything together in the present, trying to possess all the spaces of power and of self-assertion; it is to crystallize processes and presume to hold them back. Giving priority to time means being concerned about initiating processes rather than possessing spaces. Time governs spaces, illumines them and makes them links in a constantly expanding chain, with no possibility of return.
>
> *(Francis 2013, §221–222)*

On Francis's understanding, temporal processes could be compared to an underground stream, which incessantly undermines the seemingly stable foundations of space. This temporal effect is grounded in the quantitative and qualitative priority of time: on the one hand, time is greater than space because, in contrast to spatial enclosure limited by the present, temporal processes reach further and include within them the dimension of past and future – which accounts for the "fullness" of the temporal horizon. On the other hand, time "govern spaces" – temporal processes irrevocably *shape* space and establish connections between diverse spatial locations. The creative power of time identified by Francis echoes our discussion in the previous section where – in the case of Bytom – one space was *overdetermined* by a multiplicity of apocalyptic timelines; and – in the case of forests in Oregon and Japan – the time of the economy *governed* the two landscapes, by both forming and connecting them "in a chain of destruction."

The last point, however, raises the following question: can Francis's account of time, which emphasises the creative power of temporal processes, be reconciled with apocalypticism attentive to ruination, death, and destruction? Although Francis doesn't make it explicit – choosing to focus on the constitution power of time – he nonetheless acknowledges that time *undermines* space, as well as any attempts to "keep everything together in the present." Consequently, the contrast between Francis's teaching and apocalypticism is only one of emphasis: time, for Francis, would be *explicitly* generative and *implicitly* destructive, while from the point of view of apocalypticism, temporal processes would be *explicitly* destructive and only *implicitly* generative – hence the need to argue for the possibility of living in ruins when discussing apocalypticism.

Creation and destruction, however, don't exhaust the power of time. In fact, the work of time is also responsible for *keeping the present together*. We can illustrate the triple power of apocalyptic temporalities with the help of Srećko Horvat.

Horvat is interested in the destructive possibilities of *eschatological threats,* which have the power to cause irreversible damage to the life on the planet – for example the climate crisis, nuclear war, or pandemics. In fact, as Horvat suggests, contemporary eschatological threats are tipping points capable of destroying *time itself*:

> [W]hat if today, as climate crisis and the nuclear age collide, we are no longer confronted merely with the "ends of the world," but with an end that will end all other possible "ends of the world"?. . . how would we – or more precisely, who would (if there are no humans anymore) – measure time after a total event? An event in the true sense of the word "total," an event that encompasses the "whole" not only of humanity, its languages and history, but also other species and the planet, in a way – time itself.
>
> *(Horvat 2021, 121–122)*

I believe that there are two ways of interpreting Horvat's suggestion. The first one, which I assume is what Horvat *intends,* is that apocalypses today are so powerful that the destruction they wreck can undermine their own conditions of possibility: historical time. But as Stefan Skrimshire puts it:

> Prediction of the extinction of the *entire* human race are extremely speculative and furthermore they evade a much more painful likelihood that in the century to come the survival of those who can afford to protect themselves will be at the expense of those who cannot. This is a scenario that is in an important sense *already* true. Climate change has already so catastrophically changed the lives of the most vulnerable to extreme weather events that they may rightly consider that the end has come for them.
>
> *(Skrimshire 2023, 183)*

In other words, apocalypses always take place across multiple time and spaces, and so the destruction *of* history takes places *in* history. This leads me to a second reading of Horvat's claim, which I don't believe he intended. One way to represent the end of time is to imagine a cessation of temporal processes. Here, history would end not because of a total destruction of time but because of its absolute *fixity* and *stability*. Read this way, Horvat stumbles upon a third temporal power of the apocalypse – its ability to *stabilise* history, working alongside the creative and destructive aspects of the end of the world. Apocalyptic timelines would work with and against history, insofar as they are able to both *animate* and *stall* the historical processes and events. Interestingly, the original definition of tipping points quoted by Horvat to describe the ends of the world notes the stabilising power of apocalypses; this sentence, however, is omitted in Horvat's book (Horvat 2021, 28–29). Here is the definition in full:

> [T]ipping points *in general* can be defined as the point or threshold at which small quantitative changes in the system trigger a non-linear change process that is driven by system-internal feedback mechanisms and inevitably leads to a qualitatively different state of the system, which is often irreversible. This new state can be distinguished from the original by its fundamentally altered (positive and negative) state-stabilizing feedbacks.
>
> *(Milkoreit et al. 2018, 9)*

In short, the time of the apocalypse, in addition to generating and destroying spaces, would also stabilise them. Note that this intuition is implicit in Francis's characterisation of spatial enclosure in *temporal* terms – the stability of space is linked by Francis to "individual moments" and "the present." It also aligns with the ontology of contingency I proposed in Chapter 3, which recognises that the *lack* of necessity makes possible not only to change but also stasis – if everything is contingent, so is becoming.

While the idea that apocalypses maintain space is undoubtedly counter-intuitive, it can be found expressed in the concept of *permanent catastrophe*. As Walter Benjamin famously put it, that "things are 'status quo' *is* the catastrophe" (1999, 473). In fact, as Jonathon Catlin observes, it is climate apocalypse which aptly illustrates the Benjaminian thesis: the "radical view of the 'ongoingness' of catastrophe presents ecological devastation as an inherent feature of capitalist modernity, not a bug" (Catlin 2023, 55). Here, the destructive aspects of eco-apocalypse become an internal function of the world's existence, capable of generating and stabilising the status quo.

The film *Unrest* (2022), already mentioned in Chapter 3, illustrates well the power of time to maintain space. *Unrest* makes clear that *nothing happening* is ensured by temporal mechanisms and processes: in the film, the daily task of watchmakers is to assemble time-measuring machines; their work-time is itself meticulously measured by their supervisors. In addition to the factory-time, the flows of work, capital, information, politics, and, finally, life itself are organised by heterogeneous times of the municipality, the telegraph, and the railway, which – although not synchronised – conspire to hold the space together. Time also structures seemingly "free" activities: for instance, when the two protagonists Kropotkin and Josephine go for a walk, they are asked to participate in a map-making exercise, which consists of – you guessed it – *measuring time* it takes to cross a particular distance on foot.

To capture the triple power of time, we can borrow the concept of the *threshold* from the definition of tipping points above. If a threshold marks a change leading to "state-stabilising feedbacks," then the notion in effect blurs the distinction between transformation and stability. Tipping points both change and fix states; accordingly, apocalypses both alter and stabilise the world. Importantly, as the notion of permanent catastrophe makes clear, apocalypses would constitute features *internal to* the world, indicating degrees of intensity at which quantitative changes – or lack thereof – can *lead to* or *block* qualitative transformations.

In the previous section, I proposed to define the world as a threshold *in between* apocalypses. We can now see that this definition presupposes the notion of apocalypses as a threshold *internal to the world*. The earth-ground and the world-horizon can be transformed and stabilised, because the world is always already structured *from within* by temporal tipping points which creates, destroy, and fix spatial possibilities of meaning. These processes become, on the one hand, legible in landscapes, which constitute spatial memorials to the creative, destructive, and stabilising power of apocalyptic tipping points capable of shaping space, and, on the other hand, experienceable in mundane interactions with our homeworlds, insofar as the latter are transformed and maintained by the ends of the world. In short, as Lady Macbeth aptly quips, earthlings "by destruction dwell in doubtful joy" (Shakespear 1623).

The economic pluriverse

In this section, I would like to clarify the position of the capitalist timeline in relation to other apocalyptic timelines. This clarification is needed because capitalism plays two seemingly distinct roles in our examples of urban and forest landscapes. On the one hand, in the case of Bytom, the effects of post-industrial capitalism, resulting in the closing of the mine, are only *one of the many* heterogenous historical causes responsible for the shape of the home-world, alongside the timelines proper to communist planning, the Potsdam conference, individual deaths, Hussite invasions, etc. On the other hand, in the case of forests examined by Tsing, capitalism is the *dominant* timeline which produces, destroys, and connects landscapes across the planet.

The question then is this: is the capitalist apocalypse *one of many* determinates of space or *the primary cause* responsible for local ecologies of ends, to which other local causes could be reduced? This question implicates us in the old Marxist problem of economism, concerned with the status of non-economic historical causes: are the latter autonomous and effective on their own right, or are they merely epiphenomena of the economy?

In response to the challenge of economism, Louis Althusser argues that every context is structured by *dominant* and *subordinate* causes, which regularly exchange roles. Althusser critiques economism for turning the economic cause into the "eternally" dominant element, which sets up "the hierarchy of instance once and for all, assigns each its essence and role and defines the universal meaning of their relations" (Althusser 1969, 213). However, Althusser endows economy with a *special status* – he designates economic causes as *determinants in the last instance*. This epithet takes him dangerously close to economism, and, so, to avoid embracing the position he criticises, Althusser asserts that the determinant in the last instance is, strictly speaking, *absent*. "From the first moment to the last, the lonely hour of the 'last instance' never comes" (Althusser 1969, 113).

The paradoxical effectivity of an *absent cause* works in four ways. First, the economy as an *element* of a conjuncture is only one of many determinants, which may or may not be dominant. For example, in the case of Bytom, post-industrial capitalism is currently the principal determinant, though it wasn't so previously. Second, capitalism is responsible for "the permutations of the principal role between the economy, politics, theory, etc." (Althusser 1969, 213). As the *determinant in the last instance*, capitalism causes changes in the arrangements of elements in a particular conjuncture; these changes, in turn, enable us to differentiate between contexts, which are defined by the distinct composition of dominant and subordinate aspects (e.g. we can distinguish between the landscape of Bytom and the forest of Oregon because of the composition of their respective elements). Third, as the *absent cause*, capitalism "disseminates itself" in its effects – the economy doesn't determinate in the last instance by hovering over or hiding behind history; it is present *in* history as one of the elements in a given conjuncture and *as* particular history in the form of localised changes in the specific arrangements of elements. For example, in Bytom, economy *qua* absent cause is responsible for the alteration of the status of industrial capitalism, from a nascent, through a dominant, to a subordinate element shaping the area; however, it can appear only *in* and *as* this particular change and *in* and *as* a specific capitalist formation. Finally, the absent cause determinate in the last instance *connects* spaces – as we have seen in our discussion of Tsing's analysis of Oregonian and Japanese forests. However, while for Tsing this connection was established by parallel processes of ruination, and so, the two forests displayed a degree of "solidarity" in face of capitalist destruction, for the most part capitalist economy links differentiated spaces in an unequal and exploitative relation. As Pope Francis observes:

> A true "ecological debt" exists, particularly between the global north and south, connected to commercial imbalances with effects on the environment, and the disproportionate use of natural resources by certain countries over long periods of time. The export of raw materials to satisfy markets in the industrialized north has caused harm locally, as for example in mercury pollution in gold mining or sulphur dioxide pollution in copper mining . . . developing countries, where the most important reserves of the biosphere are found, continue to fuel the development of richer countries at the cost of their own present and future. The land of the southern poor is rich and mostly unpolluted, yet access to ownership of goods and resources for meeting vital needs is inhibited by a system of commercial relations and ownership which is structurally perverse.
>
> *(2015, §48–52)*[1]

If we translate this model of capitalism into the language of worlds, we can suggest the particular ecologies of ends correspond to the *many worlds* while the uneven connection between them can be designated by the notion of the *one world*. Recall that the latter is accused of being an imperialist weapon which aims to reduce the plurality of worlds. The opponents of the *one* world often quote the Zapatista slogan of "a world in which many worlds fit" (Marcos 2002, 169) – a dictum which captures the goal of "the struggle to maintain multiple worlds – the pluriverse" (Escobar 2016, 20). However, the Zapatista slogan allows also for a *dark* reversal: in an important sense, the *one* world is already *multiple*. As we have seen, capitalism – as an element within particular ecologies of ends – differentiates between worlds; however, capitalism also creates a planetary and unequal connection between these plural contexts. Capitalism, therefore, produces "a world in which many worlds fit." Consequently, the conceptual distinction between *one* and *many* worlds becomes a false dichotomy: today, the capitalist world doesn't subsume many worlds because they are different; the many worlds are different because they are produced by capitalism.

Of course, I don't mean to undermine the "ontological struggles" waged to defend the many worlds (Escobar 2016, 20). What I would like to suggest, however, is that *today*, when capitalism has become a planetary economic system, the actual line of antagonism isn't drawn *between* many worlds and the one world but *within* the uneven connection of many worlds which structures our planet. Strictly speaking, the one world can't be confronted – capitalism disseminates itself in its effect, and thus it is present only in its local and context-specific avatars populating and constituting the many worlds. Thinking globally and acting locally becomes a necessity. Thus, following the Zapatistas, we can represent capitalism as a multiple-headed hydra (2016); however, we should add that the *body* of the beast remains absent, or rather, it is found only in its multiple heads, which makes the task of killing the monster so much harder.

The ambivalence of (re-)birth

To conclude this chapter, I would like to analyse in more detail the *subjective responses* to existing in a world between apocalypses; more specifically, I would like to turn to the processes of normative and material appropriation which result in the (re-)constitution of the world after the end of the world. This discussion will offer some preliminary insights, which I will develop further in the remaining chapters of the book.

In our discussion of world-breakdowns, we identified three distinct responses to the experiences of apocalypses: carrying the world of the other, constructing narratives of redemption, and resisting through political struggle. Let's begin by taking a closer look at redemption narratives.

As Arnault points out, the English language is filled with sayings which redeem experiences of world-breakdown by ascribing meaning to them.

> Not surprisingly, given our commitment to the idea that good always arises from the wreckage of cruelty to restore meaning and purpose to our lives, moral formulas such as "Every cloud has a silver lining," "Suffering is good for the soul," and "Everything happens for a purpose" are staple forms of consolatory rhetoric.
>
> *(Arnault 2003, 156)*

In fact, the temptation of a redemption narrative may be experienced by some of this chapter's readers. Although ruins are regrettable, doesn't the fact that they create new conditions of life in an important sense *redeem* the derelict landscapes? The seduction of redemption is compounded by aesthetic experiences: there is something romantic about an abandoned forest which provide a playground for diverse species; similarly, old castles, post-industrial landscapes, or even depopulated "ghost towns" can be redeemed aesthetically as eerie spaces of possibility. The reader, therefore, may suggest that creative capacities of apocalypse, although unable to fully neutralise its destructive effects, turn the end of the world into an overall *justified* event or process. Implicit in such an intuition is a belief in the *miracle of birth*: if new life can emerge from ruins, then sites of destruction become hopeful signs of rebirth.

Interestingly, it is Derrida who picks up on this normative transformation of the end of the world, centred on the association between *death* and *birth*. When commenting on the equivocal meaning of the word "to carry," he notes that when we *carry the other* after their death, that is, when we integrate the other's world into ours, we may find ourselves confronted with new possibilities of thought and action, contributed by the dead other's norms and networks of meaning. In such a case, the response to the other's death engenders a new homeworld, and, as such, carrying the (dead) other in me becomes associated with "the experience of *carrying* a child prior to its birth" (Derrida 2005, 159). This shift of associations, and the temptation to think of death as a function of rebirth – which, at least in part, redeems the loss of the other – can be found also in responses to other types of apocalypses, where creation (of life, of worlds, of possibilities etc.) can justify the ruination and destruction which accompanies it. The attempt here is to reduce the uncomfortable ambivalence of the end of the world by showing how "the good" creation outweighs "the bad" destruction, or how "the bad" death is necessary for "the good" life to emerge.

However, as Arnault points out, redemption narratives have a direct *neutralising* impact on the material response to the end of the world, namely resistance:

> The focus on discerning the "higher purpose" of suffering – endemic to the idiom of redemption – not only implicitly calls on members of oppressed groups to accept a construction of reality that may effectively trivialize or marginalize the cruelty they are enduring, but also entices them to take satisfaction from self-sacrifice, acquiescence, and quietism.
>
> *(Arnault 2003, 181)*

My goal in the remainder of this section is to counteract the temptation to construct redemption narratives in the face of the end of the world. To do so, I will put into question the belief in the miracle of birth which silently leads many of us to redeem death and destruction by means of rebirth and creation. Admittedly, some apocalypses may involve happy endings, however, even in such instances, the ambivalence of the end of the world cannot – and should not – be reduced.

One way of putting into question redemption narratives can be found in Alain Badiou's reading of St Paul. Badiou rejects the idea that resurrection must necessarily pass through the negation of death. As Badiou points out, when the Apostle recounts his tribulations and near-death experiences, he "ascribes no redemptive signification" to his troubles (2003, 67). "Paul's preaching includes no masochistic propaganda extolling the virtues of suffering, no pathos of the crown of thorns, flagellation, oozing blood, or the gall-soaked sponge" (2003, 68). Paul's suspicion of redemption narratives is reflected, according to Badiou, in Paul's theology. The Apostle introduces a radical *disjunction* between death and resurrection, which invalidates the belief that resurrection is a sublation or an overcoming of death. The key term for Badiou is *extraction* – resurrection, although emerging within a site constituted by death, is *extracted* or *subtracted* from its relationship with death. One of the effects of separating death and resurrection is that the latter *can't redeem* the former – redemption would presuppose a relationship between death and resurrection (one of sublation or negation), however, this relation is precisely what Paul's theology implodes (2003, 71–73).

Although I am sympathetic to Badiou's reading, it is clear that the French philosopher operates with a belief in the miracle of life. The disjunction of death and resurrection results in a choice between "the bad" and "the good," with the latter being identified with life and rebirth. Consequently, even though Badiou can help us avoid justifying destruction by creation, this solution would nonetheless reduce the ambivalence of apocalypse by splitting it into two halves, "the bad" and "the good," for example by separating ruins *qua* signs of abandonment and ruins *qua* possibilities of life. However, extracting creation from destruction, or life from death, remains a type of redemption narrative, insofar as it aims to isolate the "silver lining" or the "purpose" from an essentially ambivalent site.

It is Simon de Beauvoir's account of pregnancy and birth which – in a quite literal way – can help us put into question the belief in new life as a miracle. De Beauvoir notes, proposing a critique of Badiou *avant la lettre*, that (male) attempts at dissociating life and death are symptoms of a double repression of the *materiality* and the *finitude* of life:

> This quivering jelly which is elaborated in the womb (the womb, secret and sealed like the tomb) evokes too clearly the soft viscosity of carrion for him not to turn shuddering away. Wherever life is in the making – germination, fermentation – it arouses disgust because it is made only in being destroyed; the slimy embryo begins the cycle that is completed in the putrefaction of death. Because he is horrified by needlessness and death, man feels horror at heaving been engendered.
>
> *(De Beauvoir 1972, 177–178)*

We encountered the association between the tomb and the womb already in Derrida; but, as de Beauvoir makes clear, the ambivalence of life is *not* an effect of its relationship with death; rather, generation of life is *itself* ambivalent, and it is as such that it relates to death. Childbirth is painful and dangerous, and for some women it can amount to "martyrdom" (though others may "consider the ordeal a relatively easy one to bear") (1972, 522). Furthermore, giving birth is preceded by physically taxing pregnancy (there is a "loss of appetite and vomiting" and "of phosphorus, calcium, iron," "metabolic overactivity excites the endocrine system," "blood shows a lowered specific gravity") (1972, 62). Finally, as Sabina Spielrein points out, pregnancy presupposes sex and conception, which themselves are marked by biological, psychological, and symbolic ambivalence, mixing destructive and creative elements (1994).

The ambivalent normative structure of the literal emergence of life can be applied to its more figurative counterpart. In the case of apocalypses, we cannot separate "bad" destruction from "good" creation, because the latter is always already ambivalent. Interestingly, this insight is captured by none other than St Paul, who, in the Letter to the Romans, observes that "the whole creation has been groaning as in the pains of childbirth right up to the present time" (Rom. 8:22), as well as by Marx, who likens violent force to "the midwife of every old society which is pregnant with a new one" (1976, 916). Whether we speak of *pains of childbirth* in nature, or of *force* making possible revolutionary change, we express the same intuition: creation, or the emergence of new conditions of life, is essentially *ambivalent*. Of course, we should be careful not to misinterpret these metaphors of birth as affirming, especially in the case of St Paul, some type of essentialist view of women; instead, we should read these passages as speaking to the normative ambivalence of *any* generation of life, whether that's through pregnancy, natural process, or political events.

This analysis should dispel the belief in the miracle of birth: not only is there a connection between the womb and the tomb, creation *itself* is always already ambivalent. Consequently, any narrative of redemption is in principle invalidated – because there isn't a "good" rebirth, resurrection can't redeem death, nor can generation justify destruction. Apocalypses, therefore, remain ambivalent and thus *unredeemable*. This conclusion has political implications: if, as Arnault has pointed out, narratives of redemptions neutralise resistance, then rendering them inoperative can open up a space for a disillusioned, and therefore more effective, apocalyptic politics.

A requiem

The central argument of this chapter can be *visually* summarised by the opening scene of Werner Herzog's documentary *The Fire Within: A Requiem for Katia and Maurice Krafft* (2022).[2]

Herzog's film tells the story of the life – and death – of two volcanologists, consumed by their passion for deadly pyroclastic clouds and the all-destroying flows of lava. The documentary begins with a shot of Katia standing against the fiery backdrop of lava spitting out from an active volcano; in between her and the fire, we can see a dark outline of a volcanic rock formed by the eruptions.

The opening sequence is marked by a certain confusion of times and historical associations: on the one hand, the quality of the footage, which resembles old home-made videos (indeed, Herzog uses material shot by Kraffts' themselves), is juxtaposed with Katia's protective suit reminiscent of a futurist cosmonaut; on the other hand, the red-hot lava erupting behind Katia, a violent expression of primordial geological processes, evokes the aesthetic effect of a baroque fountain, say, the one in the gardens of the Schönbrunn Palace. The temporally disjoined aspects of the shot are "held together" by the choral music reminiscent of classical requiems, as much a threnody for Kraffts as for the landscape mowed down by pyroclastic clouds. The music marks the volcanic eruption and the figure of Katia – the signs of viral nature and of life's excitement – with a poignant sense of elemental destruction, impending catastrophe, and the inevitability of death, further solidified by Herzog's voice which announces that Kraffts' lives were taken by the power of their beloved volcanos.

The scene shows a snapshot of a world constituted by apocalypses. The viewer can *see* space as an intersection of heterogeneous temporalities populated with catastrophes, which, ironically, appears frozen in time by the incessant fire which dominates the shot, giving the space an air of fixity and inescapability. This apocalyptic space embodies also the possibilities of life and death, respectively chosen and risked by the Kraffts. Everything is underwritten by an essential ambivalence – the lava destroys and forms landscapes, volcanos animate and end Kraffts' lives, the Kraffts are the embodiment of vivacity and death, and the music both exalts and condemns the world.

Admittedly, absent from the film is an explicit engagement with capitalism; yet the possibilities of life in the shadow of pyroclastic clouds and flows of lava presupposes the economic process which connects and differentiates places across the globe visited both by Kraffts and by volcanic disasters. Charitable viewers, therefore, may interpret Herzog's omission as an indication of the economy as the *absent cause*, disseminated in its effects, whose hour, as Althusser believes, never comes.

Notes

1 In this section of the encyclical, Francis also quotes a message from the Bishops of the Patagonia-Comahue region of Argentina from 2009, which echoes Tsing's insights regarding ruins left behind by capitalism: "We note that often the businesses which operate this way are multinationals. They do here what they would never

do in developed countries or the so-called first world. Generally, after ceasing their activity and withdrawing, they leave behind great human and environmental liabilities such as unemployment, abandoned towns, the depletion of natural reserves, deforestation, the impoverishment of agriculture and local stock breeding, open pits, riven hills, polluted rivers and a handful of social works which are no longer sustainable" (in Francis 2015, §51). For a further discussion of *organised abandonment* by capitalism, see Ruth Wilson Gilmore (2007).

2 I am thankful to Ozan Toksoz-Blauel for recommending Herzog's documentary to me.

Bibliography

Althusser, L. (1969) *For Marx*. Trans. B. Brewster. Harmondsworth: Penguin Books

Arnault, L.S. (2003) "Cruelty, Horror, and the Will to Redemption," *Hypatia* 18(2), pp. 155–188

Badiou, A. (2003) *Saint Paul: The Foundation of Universalism*. Trans. R. Brassier. Stanford: Stanford University Press

Benjamin, W. (1999) *Arcades Project*. Trans. H. Eiland & K. McLaughlin. Cambridge, MA: Harvard University Press

Catlin, J. (2023) "Slow Catastrophe: A Concept for the Anthropocene," in *The Environmental Apocalypse: Interdisciplinary Reflections on the Climate Crisis*, J. Kowalewski (Ed.). Abingdon: Routledge

Danowski, D., Viveiros de Castro, E. (2016) *The Ends of the World*. Trans. R. Nunes. Cambridge: Polity Press

De Beauvoir, S. (1972) *The Second Sex*. Trans. J. Cape. Harmondsworth: Penguin Books

Derrida, J. (1997) *Of Grammatology*. Trans. G.C. Spivak. Baltimore: Johns Hopkins University Press

Derrida, J. (2005) *Sovereignties in Question: The Poetics of Paul Celan*. Trans. T. Dutoit & P. Romanski. New York: Fordham University Press

Escobar, A. (2016) "Thinking-Feeling with the Earth: Territorial Struggles and the Ontological Dimension of the Epistemologies of the South," *Revista de Antropología Iberoamericana* 11(1), pp. 11–32

The Fire Within: A Requiem for Katia and Maurice Krafft. (2022) Dir. Werner Herzog. United Kingdom: Brian Leith Productions

Francis, P. (2013) *Evangelii Gaudium*. Vatican: Vatican Press

Francis, P. (2015) *Laudato Si': On Care for Our Common Home*. London: Catholic Truth Society

Gilmore, R.W. (2007) *Golden Gulag: Prisons, Surplus, Crisis, and Opposition in Globalizing California*. Berkeley: University of California Press

The Holy Bible. (2011) *New International Version*. Palmer Lake: Biblica

Horvat, S. (2021) *After the Apocalypse*. Cambridge: Polity Press

Latour, B., Stengers, I., Tsing, A., Bubandt, N. (2018) "Anthropologists Are Talking – About Capitalism, Ecology, and Apocalypse," *Ethnos: Journal of Anthropology* 83(3), pp. 587–606

Lightfoot, K.G., Panich, L.M., Schneider, T.D., Gonzalez, S.L. (2013) "European Colonialism and the Anthropocene: A View from the Pacific Coast of North America," *Anthropocene* 4, pp. 101–115

Lynch, T. (2024) "A Political Theology of the World that Ends," in *Worlds Ending. Ending Worlds*, J. Stümer & M. Dunn (Eds.). Oldenbourg: De Gruyter

Marcos, S. (2002) *Our World Is Our Weapon: Selected Writings*. London: Seven Stories Press

Marx, K. (1976) *Capital: A Critique of Political Economy, Volume 1*. Trans. B. Fowkes. London: Penguin Publishing Group

Milkoreit, M., Hodbod, J., Baggio, J., Benessaiah, K., Calderón-Contreras, R., Donges, J.F., Mathias, J.-D., Rocha, J.C., Schoon, M., Werners, S.E. (2018) "Defining Tipping Points for Social-Ecological Systems Scholarship – An Interdisciplinary Literature Review," *Environmental Research Letters* 13(3), pp. 1–12

Mitchell, W.J.T. (2002) "Introduction," in *Landscape and Power*, W.J. T Mitchell (Ed.). Chicago: University of Chicago Press

Morton, T. (2013) *Hyperobjects: Philosophy and Ecology after the End of the World.* Minneapolis: University of Minnesota Press

Rudnicki, C. (2024) "Poza efektywność, poza obowiązek: Filozofia na progu," in *Studia nad projektem „Homo sacer"*, M. Ratajczak (Ed.). Warszawa: Wydawnictwo Instytutu Filozofii i Socjologii PAN

Secret, T. (2015) *The Politics and Pedagogy of Mourning: On Responsibility in Eulogy.* London: Bloomsbury Publishing

Shakespear, W. (1623) *Macbeth.* Available online: www.gutenberg.org/files/1533/1533-h/1533-h.htm [Accessed 26/10/2024]

Skrimshire, S. (2023) "Apocalyptic Time and the Ethics of Human Extinction," in *The Environmental Apocalypse: Interdisciplinary Reflections on the Climate Crisis*, J. Kowalewski (Ed.). Abingdon: Routledge

Spielrein, S. (1994) "Destruction as a Cause of Coming into Being," *Journal of Analytical Psychology* 39, pp. 155–186

Steinbock, A.J. (1995) *Home and Beyond: Generative Phenomenology after Husserl.* Evaston: Northwestern University Press

Tsing, A.L. (2015) *The Mushroom at the End of the World: On the Possibility of Life in Capitalist Ruins.* Princeton: Princeton University Press

Unrest. (2022) Dir. Cyril Schäublin. Switzerland: Cinédokké

6

WHEN THE WORLD ENDS, I WILL MOVE TO PARIS

Anxiety, apathy, and activism

In his book devoted to apocalypses, Ernesto de Martino tells a story of a young Bernese peasant admitted to a hospital in the 1940s with "a typical schizophrenic delusion of the end of the world." The patient's condition began when he "uprooted some shrubs" in the spring and was further compounded by his father cutting down an oak tree in the autumn. The acts "started off a process of breakdown"; soon, the psychological crisis turned into a catastrophe: for the peasant, the ground began to sink – in part due to the perceived and imagined changes in the water flows as a result of uprooting the oak tree – opening up a "subterranean space, depicted as a realm of the dead," swallowing up the living, whose feet would sink with the earth. The patient tried to save people from the collapsing ground to no avail. The end of the world progressed, accompanied by "the sinking of a mountain, the flattening of the earth, the transformation of the air in a smelly blue gas" (2023, 17–18). "At the center of this catastrophe was the patient, who participated in it partly as a victim and partly as one who bore the responsibility for it" (2023, 18).

For today's reader, the case recounted by de Martino, although still undoubtedly sensational, may seem strangely mundane. Isn't the uprooting of trees, the changes to landscapes, and the gradual inhabitability of the earth precisely the *overly familiar* daily reality of the climate apocalypse? We can recognise in the fascinating psychopathological condition of the young peasant a rather everyday contemporary experience of living through the ecological end of the world; in fact, similarly to the Bernese patient, we believe that the sign of delusion today is *not* noticing the environmental breakdown. In consequence, more than a fascination with the atypical, de Martino's story evokes sympathy normally assigned to understandable reactions – if we

DOI: 10.4324/9781003348511-7

bumped into the peasant on the street tomorrow, upon hearing his story, we would nod, and before moving on to whatever errands we were running, we would congratulate him on at least trying to do something about his local ecological crisis.

The familiarity with the patient's condition can be explained in two related ways. On the one hand, the panic he experiences in response to the catastrophic changes of his environment sounds like an extreme case of *eco-anxiety* – an emotion many have felt when confronted with the apocalyptic effects of climate change. On the other hand, we have become habituated to alarmist messages about the end of the world, and so the patient's cries sound like another warning about the apocalypse – which we acknowledge filled with visible concern, before "getting on with things." Even if the world is ending, we must do shopping, go to work, or grab a drink with friends; a degree of *apathy* seems necessary to cope with the omnipresent apocalypticism.

The aim of this chapter is to explore in more depth the two affective reactions to the end of the world: eco-anxiety and apathy. I believe that both emotions belong to the category of affects which Matthew Ratcliffe calls *existential feelings*. Existential feelings "constitute a sense of how *one finds oneself in the world as a whole*" (Ratcliffe 2020, 251); more specifically, they are responsible for how we experience and appropriate possibilities afforded by the world (2020, 252). It is my contention that eco-anxiety and apathy offer two ways of revealing and relating to the practices constitutive of our gradually inhabitable homeworlds. Furthermore, as I argue, the affective responses to the end of the world not only disclose but also help to influence the specific processes responsible for eco-apocalypse – both eco-anxiety and apathy can directly inform climate activism.

This chapter's argument has three parts. First, I will examine the self-undermining process of appropriating space by polluting it, which constitutes homeworlds while rendering them inhabitable. Second, I will argue that eco-anxiety tracks the experience of normative paradoxes generated by living in the world structured by climate apocalypse; also, I will show how this paradoxical normativity can be addressed through collective action. Third, I will explore the political potential of apathy. As I will argue, apathy, or indifference to the existence of the world, constitutes an affective "break" from the ideological demand of reproduction, which can lead to new ways of re-engaging with our environments – including (re-)claiming space *otherwise*.

Overall, I aim to problematise the intuitive understanding of eco-anxiety and apathy as counterproductive from the point of view of climate politics. Rather than neutralising eco-anxiety or dismissing apathy, we should uncover the ways in which both affects can underwrite transformative engagements with the apocalyptic world.

Homeworlds and pollution

In the previous chapter, drawing on the work of Anthony Steinbock, I suggested that homeworlds are constituted through processes of normative appropriation of the possibilities afforded by the world. Shared ideas, norms, traditions etc. imply particular forms of engagement with our environment, which, when experienced as familiar, produce a sense of being at home in the world. In this section, I will show that the normative grasp of possibilities is inseparable from the processes of material appropriation of space, which, as Michel Serres argues, *takes place through pollution*. On Serres's reading, material appropriation of space is essentially self-undermining: since it requires pollution, it constitutes homeworlds only by rendering them inhabitable. Consequently, material appropriation of space complicates our normative attempts at making ourselves feel at home in the world.

How is space claimed according to Serres?

> Tigers piss on the edge of their lair. And so do lions and dogs. Like those carnivorous mammals, many animals, our cousins, *mark* their territory with their harsh, stinking urine or with their howling, while others such as finches and nightingales use sweet songs. . . . Whoever spits in the soup keeps it; no one will touch the salad or the cheese polluted in this way. To make something its own, the body knows how to leave some personal stain: sweat on a garment, saliva or feet put into a dish, waste in space, aroma, perfume, or excrement, all of them rather hard things . . . *appropriation takes place through dirt*. More precisely, what is properly one's own is dirt.
> *(2011, 1–3)*

The contemporary environmental crisis is an effect of a *planetary* extension of appropriating space by polluting it.[1] Industrial waste has expanded the work of bodily fluids of the whole species. The paradox of earth claimed by species-wide pollution is that, on the one hand, human "effluents are in are inextricably blended. *We can no longer enclose a piece of land*" (2011, 67); while, on the other hand, and in direct contradistinction with the previous point, the world of climate emergency witnesses an ongoing erection of exclusionary borders, which divide space, rendering some environments and territories inhabitable. The smell of excrements, pollution near freeways, or the noise of airplane engines claim space in a way which "excludes my presence, my existence, my health, my breathing . . . in short, my habitat" (Serres 2011, 41). In Western countries, this experience is compounded by borders and border-policing, consciously aiming to create a "hostile environment" for refugees and immigrants, whose journeys are often necessitated by the international policies of the very same Western countries (Cowan 2021). As Serres notes, "the powerful pollute more than the poor"; pollution itself embodies a will to power the desire to expand one's space at the cost of others, resulting in "*the war of all against all*" (Serres 2011, 68). The victims of environmental pollution attest to the age-long practices of claiming space by burying the dead: a nation's homeland is delimited by the countless corpses of soldiers, enemies, and colonised people buried underneath it.[2] What Pope Francis calls "our common home" (2015), therefore, is unified by capitalism, global pollution, exclusionary environments, and violence. The material practices which enable us to establish a homeworld make it progressively more inhabitable.

This paradox has an impact on normative appropriation of homeworlds: we try to make ourselves feel at home in the world by grasping our environment through familiar ideas or norms, which, in effect, aim to domesticate material practices which estrange us from the world. However, as I will show later, the contradictions of pollution block, or at least, complicate the practices of normative appropriation. Consequently, our polluted homeworlds will make us feel *not-at-ease* in the world, which Mariana Ortega describes as ranging from "the experience of minimal ruptures of everyday practices" to "a deeper sense of not being familiar with norms, practices, and the resulting contradictory feelings about who we are given our experience in the different worlds we inhabit and whether those worlds are welcoming or threatening" (2016, 61). The two particular modalities of not being-at-ease in the world which I will explore in the next sections are the affective experiences of eco-anxiety and apathy. While the former tracks the normative blocks generated by existing in a world of climate emergency, the latter captures and transforms the sense of alienation produced by practices of pollution.

The normative paradoxes of eco-anxiety

Eco-anxiety, defined by the American Psychological Association as "a chronic fear of environmental doom" (Clayton et al. 2017, 68), has been steadily on the rise; as the BBC reported, the internet searches for climate anxiety in 2023 have increased 27 times in comparison to 2017 (Gilder 2023). In this section, I would like to contribute to the growing literature on eco-anxiety by articulating its *normative structure*, bringing together two phenomenological insights: the role of norms in the constitution of homeworlds (put forward by Steinbock) and the affective revelation of the possibilities afforded by the world (found in Ratcliffe's account of existential feelings). Furthermore, if, as I suggested in the previous section, an inhabitable world doesn't lend itself to seamless appropriation by norms, we can expect this tension to be expressed in the normativity underlying eco-anxiety. More specifically, my claim is that eco-anxiety is structured by two experiential paradoxes:

1) the *qualitative* paradox of having to negotiate between the first-, second-, and third-personal normative claims while also experiencing the state of the planet as a non-negotiable priority.

2) the *quantitative* paradox of experiencing an infinite responsibility for the planet, while being able to act only on a finite, localised scale.

To articulate the first, qualitative paradox of eco-anxiety, I will draw on the model of normative life proposed by Irene McMullin. Our engagement with the self, others, and the world consists of a plurality of normative perspectives. As McMullin demonstrates, we experience ourselves as claimed by first-personal demands, concerned with individual agency and well-being; second-personal demands, related to our responsibility for others; and third-personal demands, which emerge from our belonging to a shared world. On McMullin's account, the ability to respond to and negotiate these often-conflicting normative claims becomes a condition of flourishing existence (2018).

Where do moral and political demands generated by the climate crisis fit in this picture? Here we can propose three answers:

1) Environmental demands, insofar as the climate emergency concerns the shared world, would constitute a sub-category of third-personal claims.
2) Climate disasters affect specific others; this, in turn, makes environmental demands a second-person issue. Moreover, global warming generates first-personal claims, concerned, for example, with dietary or transport choices. Thus, it can be argued that environmental demands are a sub-category of *all* three normative domains identified by McMullin.
3) The respective claims related to the self, the other, and the shared world presuppose, as its ontological *condition of possibility*, a habitable earth. Without optimal environmental factors which make existence possible, negotiating between the three normative perspectives with the goal of achieving a flourishing life would be, in principle, unfeasible – simply because life itself, including life structured by norms, would become impossible. Consequently, environmental demands can be understood as a distinct category pertaining to the ontological ground of normative existence.

I believe that all three interpretations of environmental claims are correct. The normative demands generated by the climate apocalypse are third-personal norms, which "seep into" the categories of second- and first-personal perspectives; this accounts for the experience of having to negotiate between not only environmental demands and demands pertaining to other domains but also between first-, second-, and third-personal environmental claims.

However, these daily normative negotiations take place against the backdrop of the concern with the reproduction of the environmental conditions of existence. Insofar as this latter demand applies to the ontological possibilities of life, it is experienced as *non-negotiable*. Yet, insofar as this claim is experienced as a demand, it enters into the everyday activity of weighing different first-, second-, and third-personal claims. Consequently, the moral and political demand produced by eco-apocalypse situate us in a paradoxical normative landscape in which we find ourselves *negotiating the non-negotiable* ontological ground of existence and normativity.

I believe that eco-anxiety is the affective experience which is simultaneously generated by and disclosive of the above normative paradox. On the one hand, eco-anxiety is produced whenever our familiar attempts at responding to first-, second-, and third-personal demands encounter a demand that surreptitiously introduces a concern with ontological ground into the picture; on the other hand, as an existential feeling, eco-anxiety reveals to us this seemingly irresolvable normative tension; thus, the affect regenerates itself, solidifying itself and increasing its intensity.

The qualitative paradox of eco-anxiety is exacerbated by its quantitative counterpart – the desire to marry our *infinite* responsibility for the planet with our *finite* capacity for action. As Wayne Martin points out, we normally think of moral obligations as claims which are possible for us to fulfil. The principle of " 'ought' implies 'can'" tracks our intuitions about being responsible for things we have the power to influence and not being responsible for unachievable tasks. But, Martin argues, some moral obligations are guided by *infinite ideals*: "a norm or demand that retains its authority over us even in the face of our conviction that the norm itself is impossible for us to fulfil" (2009, 103). The principle of " 'ought' implies 'can'," therefore, must be complemented with a contrasting principle of "ought but cannot" – the latter applying to a set of normative claims which, given the kind of beings we are, can't be fulfilled but which, nonetheless, we feel obligated to respond to (Martin 2009).

Environmental claims seem to belong to the category of infinite ideals. Of course, it is possible for us to change our diet or choose more sustainable means of transportation; however, the planetary scope of the climate crisis generates an *infinite* demand, which can never be adequately fulfilled by finite – and consequently only partial – actions of moral agents, capable of acting only on a limited scale. The awareness that my actions are never enough, and that there will always be a gap between what I can do and what I ought to do, is both the cause and the object of eco-anxiety: the more I grasp the planetary scope of environmental claims, the more anxious I become; the more anxious I feel, the more I realise that the infinite environmental demands cannot be fully met by a finite agent like myself.

The normative paradoxes of eco-anxiety explain the sense of not being-at-ease in the world, characteristic of existing in a homeworld constituted by self-undermining material practices. Interestingly, however, recent psychological literature on eco-anxiety suggests that this climate emotion has two opposing effects: it can either lead to an action-paralysis, a feeling of being unable to do anything, or, on the contrary, it may function as a motivator of pro-environmental behaviour, including collective action, which diminishes the negative valence of the emotion (Coffey et al. 2021). While the normative paradoxes discussed in this section could account for the paralysing effect of eco-anxiety (it is understandable that when confronted with a demand impossible to negotiate and to fulfil, we become despondent), a further explanation is needed to reconstruct the link between eco-anxiety and activism. My hypothesis is that the connection between anxiety and action is established by the fact that collective agency found in activism promises (a partial) solution to the normative paradoxes of eco-anxiety.

From anxiety to agency

To explore how collective action can address the normative paradoxes of eco-anxiety, we must take a brief detour through the work of the early modern philosopher Baruch Spinoza. I am particularly interested in two claims made by Spinoza. First, that our power to act is increased when we join with other individuals: "if, for example, two individuals of entirely the same nature are joined to one another, they compose an individual twice as powerful as each one" (1996, 125). Second, that the increase or decrease of our power to act is associated with the feeling of *joy* or *sadness*, respectively; as Gilles Deleuze observes, "joy augments our *power of acting* and sadness diminishes it" (1988, 101). In Spinozism, the connection between collective action and affects is ensured by the *transindividual* character of emotions. In the words of Jason Read:

> Affects are . . . modulations of our collective and individual life; my hopes, fears, loves, and hates are influenced and shaped by the affects of others. . . . Even the most intimate affect is profoundly social. We never love or hate alone; the very affects that form the intimacy of our striving, our desire, are refracted by social relations.
>
> *(2024, 107)*

The Spinozist insights allow us to formulate the following hypothesis: joining with other people increases our capacity to negotiate and act on the environmental demands; this enlarged power to meet our ecological obligations is, in turn, augmented by the transindividual feelings of joy – or positive affects – which counteract negative feelings associated with eco-anxiety.

Recently, Nick Montgomery and carla bergman have applied a Spinozist lens to an analysis of contemporary activism. In *Joyful Militancy*, they show how political affinity groups, centred on friendship and mutual support, offer a space for a transformation of negative affects into joy, which is inseparable from the increase of our capacities to act and think.

Oftentimes, our sad affects are experienced as highly individualised and, therefore, as isolating; moreover, they are suppressed by our need to avoid pain and pursue happiness. However, as Montgomery and bergman argue, "joy does not come about by avoiding pain, but by struggling *amidst and through it*" with others. "To make space for collective feelings of rage, grief, or loneliness can be deeply transformative" (2017, 62) because such spaces allow us to break through the isolating veneer of our sad emotions and to begin feeling-with-others. Admittedly, expressing negative affects collectively is not a sufficient condition of increasing joy: there are spaces, including many activist spaces, in which the negative emotions are compounded, and joy is not allowed to increase. Montgomery and bergman refer to this phenomenon as "rigid radicalism":

> It is the pleasure of feeling more radical than others and the worry about not being radical enough; the sad comfort of sorting unfolding events into dead categories; the vigilant perception of errors and complicities in one-self and others; the anxious posturing on social media and the highs of being liked and the lows of being ignored; the suspicion and resentment felt in the presence of something new; the way curiosity feels naïve and condescension feels right.
>
> *(2017, 168)*

However, the *right* kind of affinity group in which one feel supported and cared for can offer the environment for a transformation of sadness into joy. This is because such spaces enable us to increase our power to act and to think, which, as Spinoza reminds us, is experienced affectively as joy. When we join with others, and more precisely, when we feel supported by those with whom we share our affects, we find ourselves capable of a whole range of acts which "seemed impossible or terrifying before" (Montgomery & bergman 2017, 31). This experience is often informed by what bergman and Montgomery call (following Spinoza) *common notions*: ideas and concepts which are "inherently experimental and collective," which augment our "capacities to remain responsive to changing situations" and "to modulate the forces of the present moment" (2017, 32).

Although McMullin doesn't employ Spinozist vocabulary, her account of *virtues* seems to resonate with the growth in the capacity to feel, act, and think identified by Montgomery and bergman. For McMullin, virtues are sets of affects, beliefs, and behaviour which together constitute "mechanisms for successfully negotiating . . . different problem areas; they are ways in which we respond well to normative claims in the face of human limitation, dependency, and weakness of various kinds" (2018, 69). The practical intelligence characteristic of virtues requires habituation of both emotions and cognition, which is inseparable from supporting relations with others, including friendships: "true friends are role models for adults. . . . By providing us with different perspectives on what good living looks like, friends can challenge the styles of being into which we have been habituated and help us become better versions of ourselves" (2018, 142). Consequently, friends *qua* role models can "inaugurate the process of becoming virtuous and continuously guide even the most well-developed moral agent in times of moral conflict and confusion" (2018, 107–8). Virtues cultivated with others, therefore, are "problem-solving stances" helping us to respond – emotionally, cognitively, and behaviourally – to the specific first-, second-, and third-personal dilemmas generated by our being-in-the-world.

The Spinozist analysis of activism proposed by bergman and Montgomery, supplemented with McMullin's account of virtue, shows us why collective agency may present itself as a solution to the normative paradoxes of eco-anxiety. First, intersubjective relationships can engender common notions and virtuous habits, which, in turn, can increase our capacity to successfully negotiate the normative demands we experience; consequently, while we shouldn't expect to eliminate the paradox of having to negotiate the non-negotiable, we can respond to it more effectively if our response is informed by common notions and virtues. In other words, while the problem may not disappear, our practical solutions can become *better*. Second, the intersubjective condition of common notions and virtue – friendships and affinity groups – can extend our ability to act; we may never be able to have a planet-wide influence, but thanks to working with people, we can transcend to boundaries of our individual agency. In other words, although we may never be able to fulfil an infinite demand, we can do *more* when we join with others. Importantly, the qualitative and quantitative shift in our agency, as Spinoza reminds us, is experienced affectively, as a decrease of sad passions – including anxiety – and an increase of joy. On this reading, the negative aspects of eco-anxiety are never eliminated, since the normative paradoxes remain unsolved; nonetheless, when eco-anxiety is shared with others it can help us to discover joy, friendship, and the possibility of an incremental growth as actors, thinkers, and practical problem-solvers.

Apathy or turning our back on the world

As I mentioned earlier, eco-anxiety might lead to empowerment and collective action, but it can also result in action-paralysis, despondency, and apathy. The latter affect would be associated with disempowerment, experienced as an emotional alienation from others and the normative demands of the shared world. What I would like to show in the remainder of this chapter, however, is that this intuition about apathy, although partly correct, obscures the political potential harboured by the experience of apathic estrangement: as the examples of *The Lion King's* (1994) characters Simba, Timon, and Pumba teach us, a temporary break from others and the normative demands of our familiar environments – what Timon refers to us the *turning of our back on the world* – can offer the necessary break helping us to reorient ourselves and to re-engage with the world more effectively. In fact, apathy can function as one of the conditions of activism required to confront the climate crisis.

According to popular opinion, ecological apathy is the affective culprit responsible for perpetuating the climate apocalypse. I think this criticism is partly right. There is something apocalyptic about apathy, since an apathetic agent *doesn't care* if the world ends. However, as I will show later, the experience of apathy is, in effect, a mode of existing in a manner *unaffected* by the ideological demand to protect the environmental conditions of life. As a result, apathy promises a *way out* of the omnipresent demand of ideology to reproduce our conditions of existence; or to be more precise, following Althusser's insight that we are never outside of ideology (2014), apathy provides the affective experience amounting to the *bracketing* of the ideological demand for reproduction. To put it figuratively, apathy places us "outside" of the world, and in so doing, it helps to liberate us from the confines of our current worldly conjuncture, opening a space for truly radical thought and action unfettered by the laws of reproduction. Naturally, indifference to the world doesn't necessarily lead to novel, effective, or even interesting political ideas and acts – sometimes, apathy leads nowhere; nonetheless, the fact that apathy *could* enable innovative forms of re-engagement with the world should not be ignored.

One of the affective modalities of apathy is boredom. As Jean-Luc Marion notes, to be bored is to simultaneously disengage from the world, to leave behind our interest in beings, and to place the world "in suspension" – boredom "strikes being in general with vanity"; it "sees all and nothing, all *as* nothing, all that is *as if* it were not" (1995, 119–20). In Marion's analysis, boredom becomes an apocalyptic experience insofar as it reveals the frailty and the impermanence of the world. Boredom "marks everything with the indication of caducity."

> Not that all disappears or falls, but all *can* fall and disappear. . . . Its present permanence is saturated with its abolition. . . . That which remains immediately becomes that which does not remain, that which holds coincides with that which is undone, all or nothing, without any difference. . . . That which is, having become caduke because struck with vanity, is as if it were not.
>
> *(1995, 127)*

At this point, a critic may point out that Marion's account of boredom is self-undermining: to experience the world as impermanent and finite is *terrifying*;

in fact, hasn't Heidegger – one of Marion's phenomenological masters – argued that the genuine awareness of finitude is inseparable from anxiety? (1978). If, as Marion claims, apathetic boredom makes us come face to face with the possibility of the destruction of all beings, then being bored would induce its own transformation into an opposing affect, such as anxiety – if only because boredom's ability to "strike being in general with vanity" would put into question our own survival.

What this objection misses is that in addition to suspending the world, apathy and boredom presuppose the experience of finding ourselves *outside* of the world, in a space independent from the universal extinction of beings. To put it simply, we can be indifferent to the existence or non-existence of the world, because – as both Freud and Husserl note – we have the ability to experience ourselves as *immortal*, and thus outside of, and unaffected by worldly finitude.

> It is indeed impossible to imagine our own death; and whenever we attempt to do so we can perceive that we are in fact still present as spectators. Hence the psycho-analytic school could venture on the assertion that at bottom no one believes in his own death, or, to put the same thing in another way, that in the unconscious every one of us is convinced of his own immortality.
>
> *(Freud 1925, 289)*

In this passage, Freud brings together two distinct insights. On the one hand, he points out how unconsciously, we *don't* believe in the reality of our own death. The psychoanalytical discovery, in the words of Alenka Zupančič, "is not that deep down . . . we know that we will die . . . it is rather that 'deep down' we are sure we will not die, we just superficially and 'rationally' accept it" (2022, 90). On the other hand, Freud makes a phenomenological point about how the thought of death – including the image of one's own demise – is always a thought or an image from *a point of view of* an unaffected subject.

Interestingly, analogous claims are made by Husserl. First, similarly to Freud, Husserl notes the irreducibility of an observing subject, *even during the end of the world*. While it "is possible that entropy will put an end to all life on earth, or that celestial bodies will crash into earth," this apocalypse would have sense only "as elimination of and in the constituting subjectivity" (1981, 230–231). In *Ideas I*, he makes an even stronger claim, namely that annihilation of the world is only a derivative modification of the stream of mental processes, because "the whole *spatiotemporal world* . . . is, *according to its sense, a merely intentional being*, thus one has the merely secondary sense of being *for* a consciousness" (1982, 112). The phenomenon of the end of the world becomes relative to consciousness, yet the ego remains independent from the apocalypse it experiences. "In so far as their respective senses are concerned, a veritable abyss yawns between consciousness and reality" (Husserl 1982, 111). In a manner reminiscent of the Cartesian cogito, which asserts its own existence, the moment it doubts everything, subjectivity reaffirms its indestructability whenever it thinks of its death or the apocalypse: the paradox here is that the very thought of the end of life requires someone to witness or narrate the apocalypse.[3]

Second, Husserl notes that the subjective experience of the flow of time – the so-called inner time-consciousness – implies the immortality of the ego, supporting the Freudian insight into unconscious denial of one's death. Our temporal experience consists of an ongoing process of anticipating the future, fulfilling expectations in the present, and retaining them as a memory (think here of listening to a favourite song: it "makes sense" because the current note is inseparable from our anticipation of future sounds and the memory of previous ones). Memory stores endless examples of fulfilled expectations – everything we remember confirms the incessant emergence and fulfilment of future expectations, because every memory presupposes the flow of time. Since at each moment subjectivity anticipates the future, Husserl concludes that "the future signifies an unending time" (2001, 469). The flow of time turns the phenomenological ego into "an eternal being in the process of becoming" (2001, 471).

> The ego lives on; it always and necessarily has its transcendental future before it; the expected element having this or that content need not occur, but a different content is there in its stead; something always takes place. And there is a forward directed "always" for me as the ego. . . . What is futural will be past after it was present, and it will coalesce with the current Now.
>
> *(2001, 470)*

Returning to Marion's account of boredom, we can begin to see how the affect can reveal the finitude of the world without contradiction – apathy, while striking being with vanity, taps into the subjective experience of our necessity (as irreducible observers of the end of the world), our immortality (both unconscious and grounded in the functioning of inner time-consciousness), and the resulting sense of our independence from the annihilation of life. Freud neatly summarises this conviction with a cynical joke in which a husband tells his wife: "if one of us two dies, I shall move to Paris" (1925, 298).

As I argued earlier in this chapter, the normal practices of world-making are self-undermining, generating a sense of material and normative estrangement from our environments. It is my contention that apathy is one of the modalities of not being-at-ease in the world, because it *mirrors* the experience of estrangement produced by practices of appropriation; however, in doubling our alienation from the world, it *transforms* this experience. In contrast to eco-anxiety, which affectively tracked the paradoxes of the world, apathy enables us to "sidestep" the contradictory and alienating processes of world-constitution – in apathy, we no longer care about establishing and maintaining homeworlds; in consequence, we find ourselves, psychologically, "outside" of the world. Of course, the apathetic subject doesn't float above the earth – being "outside" of the world is an experience which materially takes place *in* the world. Thus, to be apathetic is to embody a tension between occupying and polluting homeworlds *qua* material being and disengaging from the world through an indifference which renders inconsequential the survival of life.

Subjectivity capable of experiencing both eco-anxiety and apathy would be pulled in two directions at once, *twice*: first, we feel the need to constitute homeworlds, only to find them inhabitable and alienating; second, we become indifferent to our homeworlds, only to find ourselves firmly situated within them. If the former alienating tension was grasped in eco-anxiety, the latter one accounts for the sense of not being-at-ease in the world proper to apathy. Furthermore, it is my contention that, similarly to eco-anxiety, the experience of apathy and of its (ultimately impossible) self-estrangement from the world can inform political activism; to illustrate this claim I will now turn to two sets of examples: the Biblical characters of Cain and St Paul and the so-called disaster anarchy.

The politics of apathy? Cain, Paul, and disaster anarchy

In *Laudato Si'*, Pope Francis puts forward an environmental reading of the story of Cain – the crime against Abel, Francis tells us, is inseparable from a rupture of Cain's relationship with the earth (2015, §70). The blood of Abel is soaked by the earth; the land claimed by the corpse of his brother, accuses Cain, refuses to yield crops, and, ultimately, forces him into exile (Gen 4:10–12). The story of Cain, therefore, expresses the contradiction of world-making discussed earlier – the pollution of the land by Abel's blood cries out from the ground, estranging its inhabitants; the earth is ours, yet we cannot inhabit it.

As Jacob Taubes notes, Cain was an important figure for gnostic and apocalyptic authors. However, they reinterpret the status of Cain's exile: his alienation, rather than representing a regrettable punishment for his crime, stands for an admirable refusal of the world worthy of a cult, an essentially empowering condition fuelled and fuelling a "revolutionary hatred of the world" (Taubes 2009, 38), The reader can be reminded here of the apocalypticism of Thomas Lynch examined in Introduction and in Chapter 3 and the negation of the world found therein, which echoes the gnostic reading of Cain, or of radical social movements, which as Montgomery and bergman point out, "begin with a scream of refusal: NO, ¡*Ya Basta*!, Enough!, Fuck off. . . . One spark of refusal can lead to an upwelling of collective rage and insurrection" (Montgomery & bergman 2017, 63). In the words of Taubes, the apocalyptic refusal of the world enables siding "with those who are cast out and despised," and engaging "with all the question and afflictions which are *left* by the established order of the world" (2009, 39).

The shift in the interpretation of Cain's alienation – from estranging punishment to empowerment – is not simply a literary conjecture or a heretical exegesis; I believe that it also figuratively illustrates a possible development internal to the experience of apathy. So far, we have seen how the world alienates us, and how this estrangement is both mirrored and overcome in apathy – to be apathetic is to be indifferent to the world, "outside" of it, seemingly unaffected by the processes of alienation found in the world. But, since the experience of apathy takes place *in* the world, my indifference can't be maintained forever, and it will sooner or later be interrupted by something (or someone) provoking my concern. This means that the escape from the contradictions of the world can only be temporary – similarly to Micheal Corleone in the third *Godfather* (1990), when we think we are out, we are pulled back in. In fact, even Cain ends up founding a city after his experience of exile (Gen 4:17).

My suggestion is that a temporary experience of apathy has *consequences* – it leaves a trace, an "aftertaste" as it were, which orients our re-re-engagement with the world. More specifically, apathy can have a pedagogical function, since it enables us to see the form of the world as essentially *impermanent* and thus capable of moulding. In boredom, the world is "struck with vanity, is *as if it were not*" – importantly, the collapse of the distinction between being and not-being is enacted in and as *subjective experience* (Marion 1995, 127). Here, instead of relying on plastic being capable of self-destruction (examined in Chapter 3), we turn to subjectivity as the seat of the world's impermanence. The alienation experienced as apathy reveals not only *that* the world could be suspended or ended but that it is us as subjects who can bring about this change – we have the power to engage with world as if everything could be otherwise or not be at all.

Certainly, the experience of apathy runs a twofold risk: first, it can lead to an absolute lack of concern, a full-blown nihilism incapable of motivating any actions; second, it can support an erroneous sense of subjective omnipotence, compounding the (false) immortality experienced in apathy. Nonetheless, in between the nihilist pessimism and the god-like over-optimism, and tactically drawing on both tendencies, we can find a politically useful experience of being *in* and *against* the world, which enables us to engage with the world *oppositionally*.

A reader familiar with the Judeo-Christian canon would undoubtedly recognise in Marion's characterisation of boredom as capable of rendering the world *as if it was not* a reference to St Paul. In the famous passage from his first letter to the Corinthains, Paul links the end of the world with living "as if":

> What I mean, brothers and sisters, is that the time is short. From now on those who have wives should live as if they do not; those who mourn, as if they did not; those who are happy, as if they were not; those who buy something, as if it were not theirs to keep; those who use the things of the world, as if not engrossed in them. For this world in its present form is passing away.
>
> *(1 Cor. 7:29–31)*

It is easy to see how this passage can be taken as attesting to a nihilistic strand of Paul's apocalypticism – in the words of Giorgio Agamben, Paul revokes and undermines our existential condition, preparing his readers for its end (2005, 24–25). It is this nihilistic tendency which allows someone like the ancient theologian Marcion to develop a gnostic reading of Paul on the basis of which rejection of creation in its entirety becomes possible (although, as Taubes notes, another explicit reason for Marcionite refusal of the world is the existence of too many mosquitoes (2004, 56)).

However, Paul himself seems to treat living "as if" as an element of his political engagement: as both Badiou and Taubes point out in their respective readings of Paul, the apostle's letters don't reveal a passive and disengaged nihilist but an active "leader of a party of faction" (Badiou 2003, 21), intervening in the lives of Christian communities and concerned with practical, worldly concerns: political legitimatisation, money, collective practices, tensions between Christian Jewish and Christian Gentile groups, etc. "Paul understands himself as outbidding Moses. . . . And his business is the same: the establishment of a people" (2003, 39–40). If the apostle recommends living *as if the world was not*, it is only to more effectively mould our existence in the world.

Interestingly, in a contemporary context more directly relevant to the question of climate politics, the conjunction of apocalypticism, exit from the world, and transformative political engagement can be found in Rhiannon Firth's account of *disaster anarchy*.

Firth notes how catastrophes are often followed by an emergence of decentralised, "anarchist-inspired mutual aid disaster relief efforts" (2022, 5). Drawing on the work of the Out of the Woods collective, Firth distinguishes between disaster *communities* and disaster *communisation*. The former are based on mutual aid but are ultimately "short-lived, apolitical, and vulnerable to co-optation" by the system. Disaster communisation, by contrast, aims to oppose both capitalism and the state and consists of "building links between different disaster communities, class struggles and social movements, recomposing public spaces and social bonds, and building temporal links between episodic disasters and the longer historical process of disaster capitalism" (2022, 87). The goal of disaster communisation is the creation of "immanent utopias as 'lifeboats'" in reaction to the threat of the societal collapse caused by climate apocalypse (2022, 87).

Importantly for our purposes, the process of effective disaster communisation has three elements. First, there is an *exodus* from the world, as "an inevitable response to collapse." Second, there is *re-composition*: "the creation of self-valorising autonomous affinity groups or communes." These – in a manner reminiscent of *Joyful Militancy* – are based around friendships and being-with-others. Third, there is *insurrection*, "encompassing subsistence practices, economic localisation, and networking among different communes, as well as militant resistance" (2022, 88).

Our account of apathy can be neatly grafted onto the tripartite structure of disaster communisation identified by Firth. The world of climate apocalypse is permanently alienating; we then mirror this experience of estrangement through an emotional disengagement – a form of an affective exile or exodus. Subsequently, in and through apathy, we experience an opportunity to re-compose by realising, on the one hand, our ability to make the world impermanent, and, on the other hand, our autonomy in relation to the ending world (manifested as the unconscious belief in our immortality). Finally, we can allow ourselves to be dragged back into the world ready to mount a counterattack by reshaping our environment. "Alienation and separation produce constant crises and disasters but . . . suspension of normalcy and control can be exploited to liberate potential for self-organisation" (Firth 2022, 88).

Admittedly, while the return from the exile of apathy may result in grasping the possibilities of the world *otherwise*, the re-engagement with our environment presupposes claiming space – introducing the risk of becoming complicit in the alienating processes of appropriation, which render our homeworlds inhabitable. To address this worry, I will conclude this chapter by sketching two alternative forms of claiming space: *tenancy* and *expropriation*.

Tenancy or expropriation?

What alternatives are there to the appropriation of the world through pollution? Serres's own suggestion is to give up on the notion of ownership of space all together; as he puts it, we should inhabit the earth as tenants, which would require freeing ourselves from practices of appropriation, resulting in a world which no longer belongs to anybody (2011, 78–80).[4]

> The world, which was properly a home, becomes a global rental, the *Hotel for Humanity*. We no longer own it, we live here as tenants . . . we

should no longer be the masters and possessors of nature. The new contract becomes a rental agreement. Once we have become mere renters we will be able to contemplate peace – peace among humans because peace with the world.

(2011, 72)

I am certain that anyone with an experience of living in rented accommodation will be intuitively suspicious (if not outright dismissive) of Serres's solution. Although renting may translate to a liberating experience of impermanence, it also presupposes a sense of disempowerment, of being at the mercy of the goodwill of the landlord. Furthermore, as Gerard Kuperus observes, Westerners *already* display a "superficial form of being at home" (2016, 6) – a type of *existential homelessness*, a lack of connection to places and environments, obscured by familiar settings of homogenous hotels or fast-food chains we find everywhere in the world. Here, nomadism thinly veils alienation, privilege, and ignorance. Finally, sometimes it becomes necessary to claim space and establish a home. As bell hooks observes:

African-American people believed that the construction of a homeplace, however fragile and tenuous (the slave hut, the wooden shack), had a radical political dimension. Despite the brutal reality of racial apartheid, of domination, one's homeplace was the one site where once could freely confront the issue of humanization, where one could resist.

(1990, 384)[5]

When it comes to the question of claiming space, we seem to be stuck between two unviable options sketched by Serres: on the one hand, pollution appropriates space and creates homeworlds by making the earth inhabitable; on the other hand, tenancy – although able to avoid some pitfall of its counterpart by rejecting appropriation – reproduces privileged Western nomadism and blocks the possibilities of establishing homes as points of resistance. The way of this dilemma would require a paradoxical capacity to occupying space without appropriating it. I propose to call this paradoxical form of space-taking *expropriation*.

Expropriation is a central economic concept of classical anarchism; Kropotkin, for example, views it as a necessary antidote to the appropriation of "the produce of others' toil" which characterises capitalism. For Kropotkin, the point of expropriation is to both redistribute property (acquired through the theft of the fruits of the worker's labour) and rearrange society in such a way that the very idea of property – and the types of alienation property implies – ceases to be operative. "We do not want to rob anyone of his coat, but we wish to give to the workers all those things the lack of which makes them fall an easy prey to the exploiter" (2015, 46).

Echoes of Kropotkin's treatment of expropriation can be found in contemporary forms of anarchism and autonomism, which advocate for a "transfer of ownership of goods, land, resources and means of production from private, capitalist hands to the commons" (Firth 2022, 87). Here, expropriation amounts to both releasing land from its enclosure and ending the property relations which make private appropriation of land possible in the first place. Expropriation can also be found operative in more mundane behaviours. As Michel de Certeau observes: "[e]veryday life invents itself by *poaching* in countless ways on the property of others" (1988, xii). In such cases, expropriation amounts to "clever tricks, knowing how to get away with things" (1988, xix). De Certeau refers to these daily tactics as bricolages, which create new possibilities from the available material. Think here of a walker who chooses not to follow a pavement or a road but instead crosses a field or climbs over a fence – the walker reclaims space, subjectively abolishing the imposed paths, engendering novel opportunities for movement. In the words of de Certeau, the walker "condemns certain places to inertia or disappearance and composes with others spatial 'turns of phrase' that are 'rare,' 'accidental' or illegitimate" (1988, 99).

In between reclaiming the commons and choosing not to walk on a pavement, we can find countless examples of expropriation – graffiti, squatting, guerilla gardening etc. – united by a paradoxical logic: expropriation *reappropriates* space but only to annul existing forms of spatial appropriation; it *occupies* space not to solidify its borders, but to rearrange spatial possibilities.

Understood this way, expropriation could oppose the material appropriation of space through pollution, without, however, giving up on claiming space altogether. If pollution creates "hostile environments," expropriation presents a possibility of *disengagement* from these practices, manifesting itself as a reclaiming of space and a redrawing of boundaries. To create a homeworld through expropriation is to (try to) dismantle the inhabitable homeworlds and the processes which sustain them. Of course, the practices of expropriation may result in even more inhabitable worlds; nonetheless, the risk seems worth taking.

Disaster utopias constitute a concrete example of claiming space through expropriation. When their homeworlds become inhabitable, through the mutual aid action, people engage in "a process of 'world making'" which has "a different relationship to space and scale than territorial nation states." The disaster utopias

> are local but, like the disasters that create the conditions for their emergence, they are unbounded by national borders; they are not exclusive to particular groups or identities, and they may connect to global struggles . . . Disasters create alternative forms of sociality and ways of being that do not require state mediation and must ultimately resist externally imposed power if they are to survive.
>
> *(Firth 2022, 85)*

Notes

1 Pollution floods the planet, producing a Biblical catastrophe of our own making: "At the global end of the formidable growth . . . you can see looming a hard representation of the Deluge: a planet completely covered with garbage and billboards, lakes saturated with waste, submarine ditches overflowing with plastics, seas covered with debris, residues, and peelings. On each mountain rock, each tree leaf, each agricultural plot of land, you have advertisements . . . the landscape is swallowed by the tsunami of signs" (Serres 2011, 69–70).

2 The violence inherent in homemaking is already reflected in ancient stories: "Romulus built the eternal city on the corpse of his brother" (Serres 2011, 7); while upon coming back to Ithaca, Odysseus "slays all the suitors of his wife as well as 12 of the maids. Welcome home!" (Kuperus 2016, 1).

3 Relatedly, as Sarah France points out, extinction narratives which attempt to escape the phenomenological need for a perceiver must grapple with their own paradox: "texts which explore extinction must be structured around or towards an ending that is, by its very definition, inaccessible – they anticipate an absolute absence that cannot be fully experienced, let alone narrated" (2023, 126).

4 Our names functions as prefigurative clues for conceptualising what it would mean to inhabit the world as tenants. "My name is Michel Serres, but it is not my own; I am just its tenant . . . there have been Michel Serreses in the past; there are today, and there will be in the future as many as there are apartments or houses to rent" (2011, 88).

5 hooks illustrates this point with her own experience of passing through a hostile white neighbourhood on the way to her grandmother's house: "those white faces on the porches staring us down with hate. Even when empty or vacant, those porches seemed to say 'danger,' 'you do not belong here,' 'you are not safe'." Consequently, the grandmother's house became a homeland in middle of a threating alienworld: "Oh! that feeling of safety, of arrival, of homecoming when we finally reached the edges of her yard" (1990, 383).

Bibliography

Agamben, G. (2005) *The Time That Remains: A Commentary on the Letter to the Romans*. Trans. P. Dailey. Stanford: Stanford University Press

Althusser, L. (2014) *On the Reproduction of Capitalism: Ideology and Ideological State Apparatuses*. Trans. G.M. Goshgarian. London: Verso

Badiou, A. (2003) *Saint Paul: The Foundation of Universalism*. Trans. R. Brassier. Stanford: Stanford University Press

Clayton, S., Manning, C.M., Krygsman, K., & Speiser, M. (2017) *Mental Health and Our Changing Climate: Impacts, Implications, and Guidance*. Washington, DC: American Psychological Association, and ecoAmerica. Available online: www.apa.org/news/press/releases/2017/03/mental-health-climate.pdf [Accessed 28/10/2024]

Coffey, Y., Bhullar, N., Durkin, J., Islam, M.S., Usher, K. (2021) "Understanding Eco-Anxiety: A Systematic Scoping Review of Current Literature and Identified Knowledge Gaps," *The Journal of Climate Change and Health* 3, pp. 1–6

Cowan, L. (2021) *Border Nation: A Story of Migration*. London: Pluto Press

Deleuze, G. (1988) *Spinoza: Practical Philosophy*. Trans. R. Hurley. San Francisco: City Lights Books

De Certeau, M. (1988) *The Practice of Everyday Life*. Trans. S. Rendall. Berkeley: University of California Press

De Martino, E. (2023) *The End of the World: Cultural Apocalypse and Transcendence*. Trans. D.L. Zinn. Chicago: The University of Chicago Press

Firth, R. (2022) *Disaster Anarchy: Mutual Aid and Radical Action*. London: Pluto Press

France, S. (2023) "Waiting for the End: Narrating and Grieving Extinction," in *The Environmental Apocalypse: Interdisciplinary Reflections on the Climate Crisis*, J. Kowalewski (Ed.). Abingdon: Routledge

Francis, P. (2015) *Laudato Si': On Care for Our Common Home*. London: Catholic Truth Society

Freud, S. (1925) "Thoughts for the Times on War and Death," in *Standard Edition of the Complete Psychological Works of Sigmund Freud*. Trans. E.C. Mayne. London: Hogarth Press

Gilder, L. (2023) "Climate Change: Rise in Google Searches Around 'Anxiety'," *BBC News*. Available online: www.bbc.co.uk/news/science-environment-67473829 [Accessed 28/10/2024]

The Godfather Part III. (1990) Dir. Francis Ford Coppola. Los Angeles: Paramount Pictures

Heidegger, M. (1978) *Being and Time*. Trans. J. Macquarrie & E. Robinson. Hoboken: Wiley-Blackwell

The Holy Bible. (2011) *New International Version*. Palmer Lake: Biblica

hooks, b. (1990) *Yearning: Race, Gender, and Cultural Politics*. Boston: South End Press

Husserl, E. (1981) *Husserl: Shorter Works*. Trans. F. Kersten. Notre Dame: University of Notre Dame Press

Husserl, E. (1982) *Ideas Pertaining to a Pure Phenomenology and to a Phenomenological Philosophy: First Book*. Trans. F. Kersten. Dordrecht: Kluwer Academic Publishers

Husserl, E. (2001) *Analyses Concerning Passive and Active Synthesis: Lectures on Transcendental Logic*. Trans. A.J. Steinbock. Dordrecht: Kluwer Academic Publishers

Kropotkin, P. (2015) *The Conquest of Bread*. London: Penguin Books

Kuperus, G. (2016) *Ecopolitical Homelessness: Defining Place in an Unsettled World*. Abingdon: Routledge

The Lion King. (1994) Dir. Roger Allers and Rob Minkoff. Burbank: Walt Disney Feature Animation

Marion, J.-L. (1995) *God Without Being: Hors-Texte*. Trans. T.A. Carlson. Chicago: The University of Chicago Press

Martin, W. (2009) "Ought but Cannot," *Proceedings of the Aristotelian Society* 109, pp. 103–128

McMullin, I. (2018) *Existential Flourishing: A Phenomenology of the Virtues*. Cambridge: Cambridge University Press

Montgomery, N., Bergman, C. (2017) *Joyful Militancy: Building Resistance in Toxic Times*. Edinburgh: AK Press

Ortega, M. (2016) *In-Between: Latina Feminist Phenomenology, Multiplicity, and the Self*. Albany: SUNY Press

Ratcliffe, M. (2020) "Existential Feelings," in *Routledge Handbook of Phenomenology of Emotion*, T. Szanto & H. Landweer (Eds.). London: Routledge

Read, J. (2024) *The Double Shift: Spinoza and Marx on the Politics of Work*. London: Verso

Serres, M. (2011) *Malfeasance: Appropriation through Pollution?* Trans. A-M. Feenberg-Dibon. Stanford: Stanford University Press

Spinoza, B. (1996) *Ethics*. Trans. E. Curley. London: Penguin Books

Taubes, J. (2004) *The Political Theology of Paul*. Trans. D. Hollander. Stanford: Stanford University Press

Taubes, J. (2009) *Occidental Eschatology*. Trans. D. Ratmoko. Stanford: Stanford University Press

Zupančič, A. (2022) "Perverse Disavowal and the Rhetoric of the End," *Filozofski Vestnik* 43(2), pp. 89–10

7

ANTINOMIANISM AND
SPECTRAL LAWS

In 2022, Europe was swept by a wave of climate protests consisting of throwing various things at famous artworks. Cake was smeared on Da Vinci's *Mona Lisa*, soup was poured at Van Gogh's *Sunflowers*, black liquid and red substance ended up on, respectively, Klimt's *Death and Life* and Vermeer's *Girl With a Pearl Earring* – to name just a few examples (Benzine 2022). The response from European officials was to sympathise with the protestors' concerns but to condemn targeting art. While the spokesperson of Mauritshuis museum in the Hague succinctly quipped that "[a]rt is defenceless" (AP 2022a), Austria's culture minister pointed out that "[a]rt and culture are allies in the fight against climate catastrophe, not adversaries" (AP 2022b).

Despite the fact that no real damage was done to the artworks, since most (if not all) of them hang behind protective screens, the legal apparatus reacted swiftly: following detention, the Da Vinci activist was sent to a police psychiatry unit (Britton & AP 2022), the van Gogh's protestors were "arrested for criminal damage and aggravated trespass" (Harrison 2022), and the Vermeer activists received two-month prison sentences – admittedly, accompanied by a comforting statement from the judge, who assured everyone that the ruling was not intended to discourage protest (AP & Reuters 2022). In this way, the environmentally conscious iconoclasm has joined the ever-increasing lists of climate actions deemed to be illegal: from disrupting public events, through motorway blockages and occupation of power stations, to damaging private property.

As the co-founder of Extinction Rebellion Roger Hallam explains, sometimes "[y]ou have to break the law." The turn to illegal or *antinomian* tactics

DOI: 10.4324/9781003348511-8

"creates the social tension and the public drama which are vital to create change."

> Breaking the rules gets you attention and shows the public and the elite that you are serious and unafraid. It creates the necessary material disruption and economic cost which forces the elites to sit up and take notice.
>
> *(2019, 101)*

Ecological antinomianism has many analogues in recent political history. Abolitionists, suffragettes, decolonial freedom fighters, civil rights and anti-apartheid activists, Andreas Malm observes, all successfully resorted to actions considered to be illegal (2020). However, what distinguishes eco-antinomians from their relatively recent historical predecessors is framing the need for illegal tactics as proportionate to the increasing reality of climate apocalypse. The more our environmental condition worsens, the more anti-nomianism becomes a necessity. As Malm famously puts it, it is "better to die blowing up a pipeline than to burn impassively" (2020, 151). Consequently, perhaps a more accurate historical analogue for Western environmentalism can be found in the slightly more distant history of religious movements, where the link between antinomian politics and apocalypticism – often accompanied by a concern for creation – is found established more explicitly.

Before being dispersed by Cromwell's forces, the early modern Diggers occupied and tilled land in "a symbolic and practical gesture intended to mark a radical opposition to the private appropriation and enclosure of common lands" (Smith 2000, 128). The medieval Hussites, in addition to preaching the end of the world and the suppression of laws, targeted sacred images in a manner reminiscent of the iconoclasm of today's museum activists (Bartlová 2009). But perhaps the most famous, and the most influential, expression of apocalyptic antinomianism can be found in the letters of St Paul. The apostle's apocalyptic eschatology, informing his engagement with newly founded Christian groups, as well as with his wider religious and political context, relies on an elaborate critique of the law – notably, in the "Letter to the Romans" – which is underwritten by a sensitivity to creation (as Jacob Taubes remarks, Paul treats natures as an "eschatological category" even though he has "never seen a tree in his life" (2004, 74)).

In parallel to the development of *practical* forms of illegality in Western environmentalism, we can also note a growth in the *theoretical* interest in antinomianism found among so-called continental philosophers, often associated with postmodernism. Issac Nevo calls this intellectual trend *metaphysical antinomianism*, characterised by "the spirit of turning the principled subversion of order into an end in itself . . . antithetical to the very ideas of unity, identity, law, form, etc." (1995, 297). In analogy to its practical counterpart, metaphysical antinomianism is also situated against the back-drop of an apocalyptic preoccupation with ends: the end of ideologies, the end of history, the end of grand narratives, the end of modernity, etc. As Motoko Tanaka observes, in the postmodern apocalypse the "sense of ending recurs everywhere . . . apocalypse still survives and is used to designate the postmodern mood" (Tanaka 2014, 19–20).

The antinomian-apocalyptic tendency found among continental philosophers can also explain the revival in the interest in the political theology of St Paul, whose writings, as Agata Bielik-Robson notes, serve as a springboard for elaborating messianic and antinomian philosophies of destruction.[1] For the contemporary Paulinian theoreticians,

> the impulse to save the world is inextricably bound with the apocalyptic impulse to destroy it, i.e., as in Taubes' description, to "let it go down" in its flawed present form and reveal a dimension of a "sacred void" from which one could begin totally anew, *ex nihilo*. The law figures here merely as a "retainer'," that is to say as the structuring factor of reality that preserves its illusory status.
>
> *(2014, 214)*

Even though the apostle Paul can be identified as the common ancestor of both environmental and metaphysical antinomianisms, I don't believe that the shared heritage implies a *direct relationship* between the developments in the respective spheres of politics and philosophy. While I have no doubt that climate activists read philosophy, or that philosophers have ambitions to write books relevant for activists, the particular histories of illegal *practice* and antinomian *theory* are relatively autonomous – the self-understanding of environmental protestors can't be explained through a philosophical reading of Paul, nor can a critical analysis of postmodern apocalypticism function as a substitute for a sociological study of climate activism.

Nonetheless, my contention is that theory offers a *necessary detour* for the understanding of ecological antinomianism. More specifically, as I hope to show in this chapter, philosophy possess the tools to identify the normative problems which haunt any attempts to escape the law, to explicate the logic of antinomianism, and to clarify the connection between illegality and apocalypticism. Consequently, philosophical examination of the presuppositions, structure, and implications of illegality can help us shed an indirect, theoretical light on why antinomianism may impose itself as a solution to the problems confronted practically by ecological activists.

I begin this chapter by considering two objections to antinomianism: first, that illegality is devoid of any recognisable normative meaning, and as such it makes no sense; second, that illegality is always already anticipated and recuperated by legal systems, which renders antinomian attempts to escape the law futile. To address these criticisms, I argue that antinomianism possesses its own distinctive normativity, which I call *spectral norms*, capable of challenging legal systems by undermining their foundations. I conclude this chapter by exploring the relationship between spectral antinomianism and the triple power of the apocalypse to destroy, create, and maintain worlds – a relationship which turns illegality into attractive proposition for radical environmentalism.

Two problems with antinomianism

On the one hand, throwing cake on *Mona Lisa*, disrupting sporting events, or blocking motorways may be dismissed as *senseless*. Even if we are sympathetic to the cause of climate activists, their antinomianism can appear to us as devoid of any recognisable normative meaning, situating these types of protest somewhere in between an incomprehensible avant-garde performance and acts of madness. On the other hand, eco-antinomianism, while being understandable, can be criticised for being *counterproductive*. Climate protestors are making life difficult for museum staff, for sport fans, or for workers needing to commute to earn the living; in effect, instead of putting pressure on powers-that-be and winning popular support, eco-antinomian actions are only capable of further alienating the public.

The normative presuppositions of the two common-sense criticisms of eco-antinomianism sketched here can be further articulated with the help of political theology. Carl Schmitt has famously pointed out the analogy between the exception to law in politics and the notion of a miracle in theology (2005, 36). Schmitt's intuition proves correct in our case – as I will show in this section, the two common-sense objections to antinomianism operate with distinct models of (il)legal exception which can be mapped onto two theories of miracles.

On one reading, found in David Hume's *An Enquiry Concerning Human Understanding*, a "miracle is a violation of the laws of nature" (1902). Analogously, we can think of antinomianism as a *violation* of a given legal order. Now, antinomianism could violate the law in two related ways. First, to place itself outside the law, antinomian idea or action forcefully opens up a space which is exterior to the biding legal precepts. Here violation is synonymous with an extraction or a tearing away from the law. Second, by positioning itself in a space exterior to law, antinomianism threatens the legal order by actively challenging it from the outside. Here the violation is equivalent to a confrontation between the law and its outside.

One of the consequences of antinomianism conceived of as violation of the law is that it appears *normatively empty*: if antinomianism is to be found outside the law, then its content has to be emptied of norms. This suggests that the antagonism between the law and antinomianism (or between the *inside* and the *outside* of legality) is a conflict between normativity and non-normativity or between normative fullness and normative emptiness. This could explain why, as Bielik-Robson points out earlier, antinomian philosophers can conceptualise the possibility of beginning *ex nihilo* – the nothingness from which they want to start is secured by the normative void of antinomianism. However, if we follow this reading, antinomianism will appear senseless and excessively nihilistic, since it constitutively lacks the normative principles necessary to orient any positive political or theoretical project.

We can contrast the Humean definition of miracles with the one proposed by Franz Rosenzweig. "In its time," Rosenzweig writes, "miracle demonstrated just that on which its credibility today seems to be coming to grief: the predestined lawfulness of the world." As Rosenzweig notes, "we fail to see that, for the consciousness of erstwhile humanity, miracle was based on . . . its having been predicted, not on its deviation from the course of nature" (1972, 94–95). If we approach illegality through the lens offered by Rosenzwieg, we can suggest that antinomianism, far from constituting a violent exception to the law, is rather a *confirmation* of the binding lawfulness which orders the world around us. This is because law anticipates its own transgression; this prediction, in turn, inscribes the antinomian moment within the legal system. In other words, antinomianism operates *inside* and not outside the law. Consequently, the nihilistic radicalism of antinomianism is only apparent: the seeming normative emptiness (fascinating to an antinomian philosopher and triggering to a critic of direct action) is merely a predictable function of the legal order. Antinomianism, therefore, is counterproductive and misguided, because it *affirms,* instead of challenging, the all-encompassing power of lawful normativity.

Interestingly, a clue leading to an answer for both criticisms of antinomianism can be found in Leon Trotsky's biography and, specifically, in his account of the 1905 revolution. According to Trotsky, (conservative) onlookers thought of the revolution as the instance of "collective madness"; however, Trotsky continues, revolution "appears as utter madness only to those whom it sweeps aside and overthrows. To us it was different." For the revolutionaries, the events of 1905 enabled them "to fall in love, to make new friends and actually to visit revolutionary theatres" (2007, 178–179). *From the point of view of the revolutionaries*, the chaos of 1905 was able to organise new forms of everyday life, and in so doing, to effectively oppose the old world reliant on the existing legal apparatus: in "the confusion of a revolution, a new order begins to take shape instantly" (2007, 179). Trotsky's account offers an important insight into the internal logic of antinomianism. My contention is that illegality operates with a distinctive type of norms, capable of simultaneously deconstructing the existing law and engendering new worlds. Consequently, antinomianism is neither senseless, since it possesses its own normative meaning, nor counterproductive, since it has the capacity to successfully confront the power of the law.

This chapter's argument will have the following structure: I will begin by examining the attempts to remedy the apparent normative emptiness of antinomianism by relating it to religious and natural laws. This move will prove to be problematic, as it results in affirming the inescapability of the law. To address this issue, I propose to understand antinomian normativity not in terms of religion or nature but in terms of *ghosts* with a power to deconstruct legal foundations and to create new normative environments.

(Antinomian) law against law

Bielik-Robson points out that philosophical antinomianism, in its opposition to law *as such*, reduces the differences between distinct types of law: "a prehistorical 'law of the earth', the revealed law of Torah, or an abstract legal law, appear as equally contaminated by the Fall" (2014, 215). In consequence, due to this generalisation, the antinomian messianism of destruction throws the baby out with the bathwater – antinomian philosophers, by critiquing law *as such*, miss the redemptive potential of some types of laws, which can be effectively used to oppose another type of (objectionable) legal systems. Furthermore, by programmatically aiming to start *ex nihilo*, messianism of destruction runs the risk of unleashing the uncontrollable – because extra-legal and a-normative – forces of the apocalypse, "where divine intervention cannot indeed be distinguished from sheer senseless violence" (2014, 218). On this reading, the antinomian omission of productive legal resources, therefore, can lead to an irresponsible stocking of apocalyptic fires.

In response to the perceived drawbacks of antinomian philosophies, Bielik-Robson proposes "a 'messianism of reparation'" which "treats the law not as an enemy but as its ally, however ambiguously it may be conceived" (2014, 216). The instructive Biblical theme here is not the apocalypse but the exodus:

> The event of the exodus marks the turning point in the history of redemption when the law of the Torah opposes itself violently to the law of nature and – by contrast with the latter's universal rule-without-exception, exemplified by the all-levelling heavy hand of fate – acquires militant features: it becomes – *prima facie* quite paradoxically – antinomian in regard to the natural (or, in Benjaminian idiom, mythical) laws.
>
> *(2014)*

Bielik-Robson's argument relies on two complementary steps. First, we split the genus law into distinct species (e.g. natural law, religious law, state law). Second, we allow one set of (religious) commandments to harness the destructive antinomian-apocalyptic force and to set it to work, in a domesticated and responsible form, against other forms of law. It is this "partial neutralization that allows apocalyptic energy contained within the Law to be more precise in the act of targeting its enemy"; rather than aiming to destroy the world with all its laws, commandments fuelled by antinomianism become "a more subtle missile" which targets only specific aspects of being, organised by unjust legal norms (2014, 228–229). This "operative antinomianism" can then gradually "med, fix and repair" the environment (2014, 229), functioning as the engine of "the historical process of world's emancipation" (2022, 67). "Instead of playing with fire," Bielik-Robson writes regarding messianism of reparation, "it makes fire *work* in the constant struggle with the ever-renewing powers of 'what is'" (2014, 218).

An analogous intuition can be found expressed by James R. Martel. Similarly to Bielik-Robson, Martel argues for the antinomian potential of religious law; however, as Martel demonstrates drawing on the works of Walter Benjamin, it is only one of the precepts of the Torah which can effectively harness the antinomian force: "the Second Commandment against idolatry" (2014, 22). This commandment, therefore, is both iconoclastic and antinomian – it targets "idols," which may include legal precepts, systems, and apparatuses.

> [T]his one law is the only law that we need. This law protects us from the ossification of law as a whole. . . . With it we are given just what is required for law to function as a way to safeguard and guide the human polity and no more. Indeed, as I will go on to argue, for Benjamin, by extension, all other laws must not only be superfluous but actually distort and interfere with this one, uniquely antifetishistic law.
>
> *(2014, 23)*

Thus, for Martel, just like for Bielik-Robson, antinomianism can be found *within* religious commandment(s), and in this attenuated or domesticated form, antinomianism has the power to oppose and deconstruct other laws (2014, 37).

The analogue of the alliance between religious commandments and antinomianism can be found in projects which find antinomian force in the laws of nature. Here, I would like to focus on one literary example of the latter – the essay "Law and Authority" by the anarchist Pyotr Kropotkin. On my reading, Kropotkin's text presents two interwoven arguments. First, the essay advances what may be called a "genealogical argument," which traces the development of law from uncodified social habits to contemporary legal systems. Importantly, for Kropotkin, law subsumes, utilises, and contaminates our social instincts by combing them with rules aimed to support hierarchies, power relations, and domination (1976, 33). Second, Kropotkin offers what can be referred to as a "functionalist argument," where he considers the three functions of contemporary law: the protection of property, the protection of government, and the protection of persons. Overall, Kropotkin hopes to demonstrate "the uselessness and hurtfulness of law" (1976, 39) – or more precisely, of *state* laws.

Arguably, Kropotkin is at his least convincing when he challenges the state law which aims to protect the people. In response to an objection from an imaginary interlocutor, who states that "there will always be brutes who will attempt the lives of their fellow citizens," Kropotkin writes:

> [T]he severity of punishment does not diminish the amount of crime. . . . But if the harvest is good and provisions are at an obtainable price, and when the sun shines, men, lighter hearted and less miserable than usual, do not give way to gloomy passions, do not from trivial motives, plunge a knife into the bosom of a fellow creature.
>
> *(1976, 41–42)*

While we may sympathise with Kropotkin's view regarding the absence of a direct correspondence between severe punishment and the frequency of crimes, his appeal to successful harvests and pleasant weather as an effective deterrent to violence sounds, frankly, naïve.

One possible strategy to defend this part of Kropotkin's functionalist argument is to turn to his genealogical argument. There, Kropotkin emphasises the *anteriority* of social affects and habits in relation to both state laws and religious commandments:

> Without social feelings and usages, life in common would have been absolutely impossible. It is not law which has established them; *they are anterior to all law*. Neither is it religion which has ordained them; *they are anterior to all religions*. They are found among all animals living in society. They are spontaneously developed by the very nature of things, like those habits in animals which men call instincts. They spring from a process of evolution, which is useful, and, indeed necessary to keep society together in the struggle it is forced to maintain for existence.
>
> *(1976, 31–32, my emphasis)*

The concept of anteriority of our social feelings is *natural*: it is a "past" found in the history of our biological development, which the state law – in a manner akin to an overgrown parasite – subsumes and transforms for its own authoritarian ends. Consequently, one may argue that abolishing the state law would uncover and reactivate our anterior social affects; they, in turn, would furnish a set of natural norms capable of guiding our actions in a lawless society.

It is noteworthy that the reduction of state laws to its biological foundation is framed as disclosure of another type of *law*. In another work, Kropotkin equates natural sociality and mutual aid; importantly for our purposes, Kropotkin speaks of "the law of mutual aid" (2009, 21) and considers "mutual aid as a law of nature and a factor of evolution" (2009, 24). In short, Kropotkin's antinomian alternative to state *laws* is constituted by the natural *law* of sociality. Although this recourse to legal language most likely results from its intervention in debates about biological laws, it is nonetheless significant that our social instincts lend themselves to being designed as *a kind of law*, even when abstracted from its distorted appearance within the law of the state. The good weather and access to provisions, pointed out by Kropotkin, are therefore supplementary material conditions enabling us to fully realise our natural social feelings, grounded in biological laws, by turning them into habits.

In Kropotkin's thought, antinomian normativity has temporal, ontological, and legal dimensions. To oppose the state laws, we can have recourse to past norms; this past, by virtue of being natural and instinctual, is ontologically distinct from the law which prays on our pre-existing social feelings. Nevertheless, the competing species of normativity – state laws and natural laws – belongs to a common genus of the law in general. Kropotkin, therefore, formally repeats the argument found in the political theologies of Bielik-Robson and Martel. The law is split into distinct parts; antinomianism is then combined with a specific type of (natural or religious) law, and it is set to work against other legal systems, towards the horizon of emancipation.

Recruiting antinomianism in the battle of law against law can offer a theoretical model shedding light on several phenomena proper to Western environmentalism. First, the antinomian possibilities of religious law can account for why believers can feel justified in taking part in illegal protests by the precepts of their faiths. Second, the link between iconoclasm and antinomianism expressed by the second Biblical commandment can help us to understand the recent attack on artworks and repressive measures aimed to protect the various "idols" of Western society. Third, the model of law against law can explain how norms can be disengaged from the state and used to undermine the latter's parasitic legality, as well as why antinomian groups can draw on legal rules to protect their activists and to campaign for the persecution of people and companies responsible for environmental degradation.

Furthermore, the alliance between antinomianism and specific species of law allows the former to escape the charge of normative emptiness – the seemingly nihilistic force of illegality always finds itself "clothed" or embodied by the language of laws, from which it borrows its normative content; this borrowed normativity, the thinkers examined in this section maintain, allows antinomianism to more effectively target its opponent. While this conclusion addresses the first common-sense criticism of antinomianism, it does so only by affirming the second one: the standpoint of law against law ends up confirming the overarching power of legal precepts which, by incorporating legal transgression within the body of law, neutralise the antinomian force in the process.

The holy trinity of law

While I agree with Bielik-Robson's observation regarding the antinomian tendency to collapse the distinctions between different types of law, I believe that this lack of attention is at least in part justified by the shared *form* of distinct laws. It is precisely this formal unity of law which antinomianism must transgress to liberate itself from the power of legality.

The common formal structure of state, religious, and natural laws can be divided into its "vertical" and "horizontal" aspects. *Vertically*, each type of law presupposes a guarantor or a foundation: in our examples, this role would be played by the state, God, or nature. In addition to presupposing a foundation, laws point beyond themselves *horizontally* – they evoke other species of legality to support their own functioning.

The horizontal and vertical aspects of law were already identified by Thomas Aquinas. In his *Summa Theologiae*, Aquinas characterises God as a lawgiver – the guarantor of laws – who governs creation with eternal natural laws; the latter help people to devise human law which, in turn, must be supplemented by the revealed law of the commandments, capable of remedying the doubts we may experience when engaging with natural and human laws (1947). Echoes of the Thomist legal architecture can be found in Althusser's reflections on law. While in his account the state substitutes God as the foundation, the horizontal interconnections remain in place. The notion of a legal person operative in state law, for instance, is secured by a belief that humans "are free and equal *by nature*"; these naturally free being, however, have a responsibility to honour their obligations due to the religious categories of "conscience" and "duty." State law, therefore, is "a system *that cannot exist all by itself*" (2014, 68), since it presupposes reinforcement of other legal domains.

We can model the relationship between the different types of law on the theological concept of the Holy Trinity. Analogously to the three distinct divine persons (the Father, the Son, and the Holy Spirit) existing in one Godhead, laws founded on the state, nature, and God, while not reducible to each other, co-constitute a single legal structure, united by a shared vertical form and intersecting horizontal connections. The existence of the holy trinity of law, in turn, limits the impact of antinomian elements found in specific laws. Although antinomianism embodied in laws can fight other laws, it does so only as a function of the tripartite unity of legal systems. Thus, while it is possible for one domain of law to draw on antinomianism to oppose its counterpart, such a conflict would constitute only a localised clash taking place *within* the unitary legal field, which neutralises the destructive power of antinomianism.

The hypothesis which I will explore in the remainder of this chapter is that injecting antinomianism with a force capable of challenging the holy trinity of law requires articulating illegality in relation to the figure of the *ghosts*, whose spectral normativity endows antinomianism with a power to deconstruct the holy trinity of law.

Enter the ghosts

Our analysis so far suggests that antinomianism is confronted with the following choice: it either roams freely without a content, unleashing the nihilistic fires of the apocalypse (in line with what Bielik-Robson calls the messianism of destruction), or – like a genie in a bottle – antinomian negativity is captured by a foreign medium of law; by being given a legal body, its apocalyptic force is attenuated and re-employed as a fuel driving changes in the world.

My claim is that the two options highlighted here don't exhaust the possibilities available to antinomianism. In fact, it is Bielik-Robson who, by intuiting the *ghostly* character of antinomianism, provides a clue for a third scenario. She refers to the "antinomian spectre" (2014, 15) which "hovers over the world" (2014, 196). Admittedly, Bielik-Robson reduces the ghosts of illegality to "the halo of pure negativity" (2014, 190), necessitating its embodiment in the body of law; on my reading, however, the antinomian spectres possess a distinct *positive* content, which endows it with a degree of autonomy in relation to the law it aims to oppose.

The *positive* ghosts of antinomianism can be found summoned in the pamphlet "Be Gay Do Crime." This document written by an anarchist collective the Mary Nardini Gang (MNG) offers a poetic phenomenology of queer anarchism in 21st-century United States. Similarly to Kropotkin, MNG expresses a radical antinomian position.[2] However, for MNG the source of antinomian normativity is to be found not in natural laws but in a relationship with *ghosts* of the ancestors. Here ancestral relationships are equivalent to an imaginative identification with chosen spirits rather than a biological connection – a felt relation with "a long and varied line of rioters, thieves, writers, hustlers, mystics, ranters, freaks, and artists" (2019). Importantly for our purposes, the ghosts of the ancestors generate normative content capable of directing the destructive force of antinomianism:

> We live in a world haunted by all the ghosts of a genocidal leviathan, where the land is full of bones screaming out for vengeance and the very architecture of these cities filled by all the dead who built it. . . . We received and

will do our best to transmit this mystery: queer criminality. . . . Nothing began with us: we are simply the present bearers of that weak messianic potential to make it whole again, to redeem all our dead, by way of heaven on earth . . . the ancestors have returned and they are insisting that we have a chance to make it all whole again.

(MNG 2019)

The poetic style of the MNG's pamphlet – as well as the thinly veiled references to the work of Walter Benjamin – produces a text which is undoubtedly appealing rhetorically. However, the pamphlet remains underdeveloped philosophically. To expand on the insights expressed by MNG, in the following sections I will analyse the different modalities of our relationship with ghosts and the specific character of spectral norms.

Being with ghosts

In Chapter 5, I discussed the experience of world-collapse associated with the other's death. As we have seen, for Derrida, when the other is gone, I experience the demand to "carry the other in me" (2005, 160). Drawing on Derrida's engagement with psychoanalysis, Timothy Secret discusses three ways in which we can carry – or, in Secret's parlance, "eat" – the dead: introjection, incorporation, and anti-incorporation. Each of these three digestive processes (which, as Secret argues, are intertwined, always contaminating each other (2015, 174–175)) results in *spectralising* the other, that is turning the dead person into a ghost.

Introjection refers to a healthy, albeit often painful, process of digesting the death of the other. To put it simply, I work through what the other and the other's death meant to me; here, linguistic expressions help "to clarify, correct and slowly accept" the meaning of the other's death, reintegrating this event into our lives (2015, 155). Incorporation, by contrast, marks a breakdown of healthy introjection. To put it figuratively, rather than aid the digestion of the dead, it keeps the other unprocessed "inside the stomach." Introjection "establishes a 'crypt' around the object in the psyche . . . preserving the dead intact along with the patterns of secret pleasure derived from them" (2015, 163). The dead other is buried, kept untouched, and as such sealed within the psyche. Here, grief becomes unworkable, because the other's death becomes a "forbidden" or a "magic" word, excluded from our language (2015, 163). Since the meaning of the other's death can't be brought to presence in language and exercised through linguistic expression, the other continues to haunt us *from within*. Anti-incorporation can be positioned in between introjection and incorporation. The other is neither digested nor buried in our stomachs; instead, in anti-incorporation we "keep the dead in a demarcated 'outside' rather than allowing them to haunt the psyche" (2015,

164). The example given by Secret here is Freud's locket with a picture of his daughter Sophia: the object preserves the dead other, who now haunts us *from outside*.

We should also note that it is possible to spectralise others even when they are alive and well. Think here of the phrase "they are dead to me" applicable only to living others, or of a haunting memory of a critical teacher, who – while probably relaxing at their retirement home abroad – still regularly stokes up our insecurities. In the words of Secret, "we always engage with any other, living or dead, through an economy of crypts" (2015, 173).

Furthermore, as Secret argues, the others we "eat" and turn into ghosts are haunted by their own spectres: "every crypt contains phantoms, each with its own phantoms, and so on" (Secret 2015, 169). Consequently, being with ghosts presupposes a spectral web of references, where our ghosts point to their spectres, who point to their dead, and so on. In short, we are always in contact with more than one phantom (Derrida 1994, 24).

Secret's analysis of being with ghosts can be extended beyond the relationship with *personal* others. Anti-incorporation offers a clue here: not only lockets and photographs but also books and films – among other objects – can mediate our relationship to both personal and impersonal pasts. Take engaging with a historical source as an example. As Ethan Kleinberg observes, the study of history is haunted by "the ghosts of things forgotten, buried, or erased from the record" (2017, 9). While some of the historical spectres refer to characters, others relate us to impersonal *dead possibilities* – failed uprising, supressed ideologies, missed encounters etc.

We can see clearly why spectres lend themselves so well to deconstructive putting into question of binary oppositions. Not only are they "neither living nor dead, present nor absent" (Derrida 1994, 63) – the ghosts also ignore the distinction between inside and outside, language and silence, subjects and objects, one and many, personal and impersonal. Nevertheless, spectral apparitions are united by their specific normative effects, which align them with antinomianism.

Spectral norms

Oftentimes, the appearance of a ghost is accompanied by an experience of a normative demand. All of a sudden you're visited by an image of your old university lecturer, and you reluctantly decide to rewrite a section of your paper; you catch a glimpse of an old family photograph, and unexplainedly you begin to feel shame about your current job; you come across an obscure historical event, and you become overcome with an urge to uncover its causes, protagonists, and effects. The various demands placed on us by our ghosts, Derrida tells us, share a structure: spectral normativity produces a call to *inherit*, which involves taking responsibility for the spectre

(1994, 114). But, if we are always surrounded by more than ghosts – which means that we are always haunted by more than one call to inheritance – our response must be "critical, selective and filtering" (1994, 114). In the words of Nicole Pepperell:

> Inheritance does not take the form of a passive, faithful reception of contents transmitted in pure and undistorted form. . . . Inheritance is instead an active, performative act. We – the heirs – initiate it; we select those aspects of the past that shall remain relevant to our time. . . . Our present is haunted precisely because we face a choice among multiple pasts we could inherit.
>
> *(2014, 45)*

Selective inheritance is *ambivalent* for two reasons. First, it presupposes a choice to ignore, and thus to sacrifice, other ghosts, who also demand justice.[3] Consequently, heeding to one spectral demand can become inseparable from the experience of accusation by another phantom; the ghost "does not always mark the moment of a generous apparition or a friendly vision; it can signify strict inspection or violent search, consequent persecution, implacable *concatenation*" (Derrida 1994, 126). Second, the guilty conscience generated by selective inheritance is compounded by the crimes of the ghosts we inherit: beloved family members hide dark secrets, heroes of resistance are capable of atrocities etc.

The ghostly demand is also *ambiguous* – as Hans Ruin observes, "we can never really *know* what we owe the dead or what they demand from us." The spectres confront us with the "questions of *justice* and of *obligation*" in a manner which can't be translated into exact rules, whether moral or legal (2018, 7). Derrida captures this ambiguity figuratively by evoking the effect of a visor – a helmet which hides the face from our view while producing in us an uncanny sense of being looked at (1994, 6–8).

Importantly, the ambivalence and ambiguity of spectral normativity doesn't prevent it from orienting specific ethico-political standpoints. Ghostly demands reconfigure not only our response to the past but also our relations with the present and the future. Discovering new facts about our ancestors from a photograph can be inconsequential; however, it can also transform my relation to myself, my relatives, and my community. A great example here is Antigone:

> The commitment to the no-longer living expresses a temporality in which her own actions and life occupy but a small part. Her brother and her parents are gone. The dead are *past*. But it is a past that is open to a future in the sense that she, in and through her own mortality, is also *part of that which will have been*. . . . In being led by a loyalty to *ancestors* rather than

complying with the juridical-political power of the king, Antigone stands out as an exemplary *necropolitical* heroine, in her uncompromising commitment to caring for the dead in burial.

(Ruin 2018, 3–4)

Overall, and importantly for our purposes, spectral norms stand for a type of normativity which is neither empty nor subordinated to law and which, as such, displays an organic connection with antinomian projects aiming to undermine "juridical-political power." While the call to inheritance marks the positive content of spectral normativity capable of orienting the ethics and politics of illegality, the ambiguity of the ghostly demand ensures its irreducibility to legal rules. Certainly, this is not to say that the call to inherit can't be translated into law; on the contrary, the question of inheritance is undoubtably an integral part of legal systems – think here of the will, a document which specifies who inherits and what is inherited, regulating the ambivalence of heritage. My point is, rather, that the *ambiguity* of spectral demands stands for a normative remnant which always escapes legality. The ambiguous call to inherit, by virtue of being relatively autonomous in relation to law, endows antinomian project with a content which can't be *completely* re-integrated into legal rules. At this point, we can answer the two criticisms of antinomianism with which we started this chapter: antinomianism can both make sense and avoid subordination to law, precisely because its spectral sense marks an excess which isn't fully recaptured by legality. Moreover, as I will show in the next section, the fact that spectral norms can be (partially) brought into law, can – paradoxically – further support antinomian projects.

Phantom foundations

In horror stories, spectre can appear as a formless spirit capable only of destruction; it can also be made to serve its earthly masters (usually through some magical incantation or artefacts). However, ghosts can also *possess* bodies – in such cases, the spirit becomes an agent responsible for the victim's unnatural movements, alien words, and ghastly bodily transformations.

My contention is that the normative ghosts of antinomianism, analogously to the possessing spirits found in horror stories, can't be reduced to either a force of destruction or a weapon wielded and controlled by a necromant. Rather, spectral norms have the power to take over and subordinate the foreign body of the law to itself; more specifically, ghostly normativity can *spectralise* (or turn into ghost) the foundations of law, offering support for antinomian projects. The possession of legal rules by spectral normativity, in turn, would mark the *reversal* of the embodiment of antinomianism in commandments found in the model of law against law discussed earlier – if the latter led to the attenuation of the antinomian force, the former effects a deconstruction of the holy trinity of law.

Walter Benjamin notes how the police – one of the repressive apparatuses of the state – is individuated by its "nowhere tangible, all-pervasive, *ghostly* presence in the life of civilized states." For Benjamin, the police appears when the state "can no longer guarantee through the legal system the empirical ends that it desires at any price to attain . . . the police intervene 'for security reasons' in countless cases where no clear legal situation exists" (Benjamin 1979, 286–287). One way to interpret Benjamin's remarks is to suggest that the police is able to confront anomie only by entering into or hovering over spaces of potential illegality. As a result, the police's presence oscillates between the inside and the outside the law, embodying the ambiguity characteristic of spectral normativity. To fight the phantoms of antinomianism, police itself becomes ghost-like: the repressive arm of the state can be compared to a protagonist of a horror story who, to find, fight, and neutralise spectres, must himself become a ghost.

The spectral character of the police, however, is only one expression of a much more fundamental relationship between ghosts and the state. Think here of the Tombs of the Unknown Soldier found across European states, built after World War I:

> This public memorial . . . was thereby made into a symbol and place of reverence for the citizens of the modern (increasingly) democratic nation-states. . . . At this very disenchanting historical moment in European history, it was around the sacralization of an unidentified corpse (or in some cases an empty grave, representing such a corpse) that a political community gave itself a new spiritual-political focus.
>
> *(Ruin 2018, 85)*

The veneration of ghosts by the state continues to this day. During recent celebrations of the Polish Independence Day, President Andrzej Duda explicitly connected the re-establishment of Polish statehood to the dead soldiers of World War I and the death of the insurgents who took part in the failed uprisings of 1794, 1830, and 1863.[4] It is their ghosts who founded and continue to found the Polish state. Analogous processes of spectralisation can be observed in many other states, insofar as they draw on their dead to renew or solidify their statehood.

As I suggested earlier, the state is only one of the three "persons" constitutive of the holy trinity of law. Importantly, the other two foundations of legal systems – nature and God – are also subject to spectralisation. The anonymous pamphlet "Desert" speaks of *ghost species* – forms of non-human life which have or will have become extinct: "[a] great many of the plants and animals we perceive as healthy and plentiful today are in fact relics or ghosts"

(2011). Other authors discuss *spectral landscapes*, irrevocably ruined, whose destruction continue to haunt us (Tsing et al. 2017). Nature, therefore, is not only populated by ghosts; due to the disastrous effects of climate change, it itself has become spectral. The ghostly character of God is even less ambiguous – at the end of the day, the Holy Spirit is one of the three persons of the Christian trinity! In the words of John D. Caputo, God "haunts existence, calling upon us in the night, leaving us to wonder if anyone was there . . . hauntology – the discourse on specters – taken seriously is the only way to approach what is going on in and under the name (of) 'God'" (2022, 3). The peculiar status of saints in Christianity is also noteworthy – they are phantoms that watch over us, intercede on our behalf, and mediate between us and God.

I believe that the possession of the foundations of law by spectral norms can aid antinomian projects. Admittedly, the spectralisation of the state, nature, and God can't be equated with a *direct* confrontation between the forces of law and antinomianism, comparable to a spectacular battle between protestors and the police. Here, the antinomian work of spectres is subtle and barely noticeable. Ghosts insinuate themselves into the respective identities of the state, nature, and God, imperceptibly turning supposedly firm bedrocks of law into *phantom foundations.*[5] Similarly to Bruce Willis's character in *The Sixth Sense* (1999), the legal apparatus may not notice the self-undermining spectral status of its guarantor. Any attention-grabbing confrontation between law and illegality, therefore, helps to cover over the already-taking place deconstruction of the holy trinity of law – the silent victory of antinomian spectres.

The curse of (non-)reproduction

In his book on the possession of nuns at Loudun, Michel de Certeau notes that the "first phantom-like 'apparitions' took place in the convent at the same time as the last plague cases in the city are mentioned – at the end of September 1632" (2000, 13). In Loudun, the visitation by ghosts marked a moment of re-constitution of a city traumatised and fragmented by a pandemic. Interestingly, ghosts play a similar role in the speech by President Duda mentioned in the previous section where spectres are linked to the re-establishment of the Polish state after its partition in the 18th century. These two seemingly distant historical examples make apparent the double-edged effect of spectralisation: in parallel to undermining legal foundations, *ghosts play a role in the creation and maintenance of worlds*. More specifically, law structures, and mediates between, normative and material appropriation of space responsible for the constitution of homeworlds;[6] and if ghosts inhabit the body of the law, then antinomian spectres are complicit in organising normative and material possibilities of homeworlds.

Some readers may equate the world-constituting power of antinomian spectres with its revolutionary potential. "It's not a question of forming a void from which we could finally manage to catch hold of all that escapes us, but of learning to better inhabit what is there . . . affirmation is an element of attack" (Invisible Committee 2015, 79). Others, however, may approach spectral norms as ideologically suspect – the risk inherent in spectral normativity is that it can always support conservative, world-preserving projects. Think here of Simba in *The Lion King* (1994), who, inspired by the ghosts of his father Mufasa, re-establishes the "rightful" kingdom, squashing the hyenas' rebellion led by Scar. Either way, the world-creating effects of antinomian spectres suggests that the apocalyptic fantasy of beginning *ex nihilo*, professed by the proponents of the messianism of destruction, becomes unfeasible. The subversive work of ghosts can't be separated from their world-forming function. The paradox of antinomianism consists of the fact that its destructive, creative, and stabilising elements are interwoven: the more we attempt to undermine the legal grounds of our homeworld, the more we lay the foundations for the emergence of a new world. It is precisely this paradox – found also in our discussion of apocalypse in preceding chapters – which establishes the intimate connection between antinomianism and apocalypticism.

Here, we once again confront the problem of the inescapability of reproduction, which resurfaces at different points throughout this book. The analysis of (legal and worldly) non-reproduction aimed at by antinomianism and apocalypticism reveals that destruction is inseparable from reproduction. Our choice, therefore, is not between reproduction or non-reproduction but between an *emphasis* on one or the other aspect of an unbroken dyadic relationship. However, rather than seeing this contradiction as the ultimate blow to antinomian apocalypticism, we can reinterpret it as a positive condition of radical environmentalism. It is because antinomian apocalypticism, in destroying laws and worlds, *creates* and *maintains* laws and worlds, that we can consistently advocate for the end of the world motivated by environmental concerns. By placing an emphasis on non-reproduction, and by hoping for the end of our homeworld, we plant seeds for a future – however ghost-like – which reorganises our material and normative relationship with our environment.

I would like to end this book by bringing up two stories that illustrate the function of non-reproduction in the substitution of worlds. The first one is the story of Noah, recently re-read by Pope Francis, who connects the catastrophic non-reproduction necessitated by the deluge with the non-reproduction expressed in the law of the Sabbath. When the world is destroyed in a great flood, through Noah, God "gave humanity a change of a new beginning."

The biblical tradition clearly shows that this renewal entails recovering and respecting the rhythms inscribed in nature by the hand of the Creator. We see this, for example, in the law of the Sabbath. On the seventh day, God rested from all his work. He commanded Israel to set aside each seventh day as a day of rest, a *Sabbath*. Similarly, every seven years, a sabbatical year was set aside for Israel, a complete rest for the land, when sowing was forbidden and one reaped only what was necessary to live on and to feed one's household, Finally, after seven weeks of years, which is to say forty-nine years, the Jubilee was celebrated as a year of general forgiveness. . . . This law came about as an attempt to ensure balance and fairness in their relationships with others and with the land on which they lived and worked. At the same time, it was an acknowledgment that the gift of the earth with its fruits belongs to everyone. Those who tilled and kept the land were obliged to share its fruits, especially with the poor, with widows, orphans and foreigners in their midst.

(2015, §71)

The experience of the flood, and the lawlessness and non-reproduction it involved, is carried on as a normative demand for rest, a *phantom law* organising a just relationship with our natural and social environment. Non-reproduction, therefore, functions as the apocalyptic remnant which survives the end of the world and whose goal is to, paradoxically, ensure *better* forms of reproduction of our environmental and social conditions of existence. In the story of Noah, non-reproduction, experienced during the disastrous apocalypse, harbours elements capable of informing the everyday existence in post-apocalyptic worlds. "And so the day of rest . . . sheds it light on the whole week, and motivates us to greater concern for nature and the poor" (2015, §237).

The optimistic tenor of the Biblical narrative can be contrasted with the original *Planet of the Apes* (1968). There, the surviving elements of the old world – the statue of liberty buried in the sand, an old pacemaker, glasses, and a doll found in the Forbidden Zone – attest to the catastrophic aspects of non-reproduction. When the experience of the latter is carried on into the new world, rather than empowering humanity, it reduces human species to the role of mute prisoners, incapable of complex collective organisation, technological developments, scientific advancements, or even play (activities symbolised by the items in the Forbidden Zone). It is also noteworthy that the viewer – especially one suspicious of the cowboy-like demeanour of the protagonist – sides with the apes against the people, in part because of the clear connection between the plot and the possibility of humanly induced apocalypse off screen (nuclear then, environmental now). Thus, in contrast to the story of Noah, *Plantes of the Apes* shows how substitution of worlds, and the remnants of lawless non-reproduction carried on from the experience of the apocalypse, possesses a dystopian power capable of imploding our conditions of existence.

The point is to think of these contrasting notions of non-reproduction – the optimistic and the pessimistic – as two sides of the same coin. Apocalypticism of everyday life situates us in the midst of processes which destroy, create, and stabilise worlds; the opportunity presented by eco-politics, attuned to the ends of the world, lies in recognising non-reproduction as a presupposition of reproduction and lawlessness as a useful resource for action. However, it must be borne in mind that the passage through non-reproduction and antinomianism for the sake of a better world always runs a risk of creating and reproducing a worst one. Nothing captures this hopeful curse of eco-apocalypticism better than an image of a devasted work of art.

Notes

1 "Thus, in *The Political Theology of Paul*, Jacob Taubes, following Benjamin and Schmitt, advocates a full *suspension* of the law . . . Alain Badiou . . . calls for a strong political reading of the Paulian messianic gesture that, according to him, consists in a *non-dialectical* rejection of the law . . . Slavoj Žižek . . . for him the gist of the Paulian messianism is the *dialectical negation* of the law . . . And finally, Agamben, in *The Time That Remains* . . . proposes a subtler solution that consists in a *deactivation* of the law" (Bielik-Robson 2014, 214).
2 "[W]e stole away on trains with forged documents, on fraudulent flights, and in the cars of strangers who picked us up en route to one encounter after another. . . . We fought enemies minute and gargantuan on streets and in alleyways. We were there when cities were burned, buildings occupied, boutiques looted, ports blockaded, wanna-be bashers humiliated, nazis punched. We delivered an empty coffin to the doorsteps of a killer cop, threw fire into the home of a john who killed a trans woman, and more through bank windows in the name of those imprisoned for refusing a similar fate. We instigated the wildest queer riots in a generation outside the gates of summits of the global elite" (MNG 2019).
3 Derrida discusses this ethical paradox in more detail in *The Gift of Death* (1996, 67–68).
4 The speech is available online: www.youtube.com/watch?v=HUYy_G2YWAA
5 I borrow the idea of *phantom foundations* from Jan Sowa, who discusses it in his book on the early modern history of Polish statehood (2011).
6 For a discussion of the relationship between law and appropriation, see Althusser (2017, 55).

Bibliography

Althusser, L. (2014) *On the Reproduction of Capitalism: Ideology and Ideological State Apparatuses*. Trans. G.M. Goshgarian. London: Verso
Althusser, L. (2017) *Philosophy for Non-Philosophers*. Trans. G.M. Goshgarian. London: Bloomsbury Publishing
Aquinas, T. (1947) *The Summa Theologiae*. Trans. Fathers of the English Dominican Province. Available online: www.ccel.org/a/aquinas/summa/home.html [Accessed 28/11/2024]
The Associated Press. (2022a) "'Girl with a Pearl Earring' Targeted with Glue and Liquid by Climate Activists," *NBC News*. Available online: www.nbcnews.com/news/world/girl-pearl-earring-targeted-glue-liquid-climate-activists-rcna54494 [Accessed 28/11/2024]

The Associated Press. (2022b) "Climate Activists Throw Black Liquid at Gustav Klimt Paining in Vienna," *The Guardian*. Available online: www.theguardian.com/environment/2022/nov/15/climate-activists-throw-black-liquid-at-gustav-klimt-painting-in-vienna [Accessed 28/11/2024]

The Associated Press & Reuters. (2022) "Climate Activists Who Targeted Vermeer Sentenced to Prison," *DW*. Available online: www.dw.com/en/climate-activists-who-targeted-vermeer-sentenced-to-prison/a-63630123 [Accessed 28/11/2024]

Bartlová, M. (2009) "Understanding Hussite Iconoclasm," *Filosofický časopis* 57, pp. 115–126

Benjamin, W. (1979) *Reflections: Essays, Aphorisms, Autobiographical Writings*. Trans. E. Jephcott. New York: Harvest/HBJ

Benzine, V. (2022) "Here Is Every Artwork Attacked by Climate Activists This Year, From the 'Mona Lisa' to 'Girl With a Pearl Earring'," *Artnet*. Available online: https://news.artnet.com/art-world/here-is-every-artwork-attacked-by-climate-activists-this-year-from-the-mona-lisa-to-girl-with-a-pearl-earring-2200804 [Accessed 28/11/2024]

Bielik-Robson, A. (2022) "The Harnessed Lightning, or the Politics of Apocalypse: Hegel, Rosenzweig, Derrida," *Praktyka Teoretyczna* 1(43), pp. 63–92

Bielik-Robson, A. (2014) *Jewish Cryptotheologies of Late Modernity: Philosophical Marranos*. London: Routledge

Britton, B., The Associated Press. (2022) "Man in Wig Throws Cake at Mona Lisa in Climate Protest Stunt," *NBC News*. Available online: www.nbcnews.com/news/world/man-wig-throws-cake-mona-lisa-climate-protest-stunt-rcna31094 [Accessed 28/11/2024]

Caputo, J.D. (2022) *Specters of God: An Anatomy of the Apophantic Imagination*. Bloomington: Indiana University Press

De Certeau, M. (2000) *The Possession at Loudun*. Trans. M.B. Smith. Chicago: The University of Chicago Press

Derrida, J. (1994) *Spectres of Marx: The State of the Debt, the Work of Mourning and the New International*. Trans. P. Kamuf. London: Routledge

Derrida, J. (2005) *Sovereignties in Question: The Poetics of Paul Celan*. Trans. T. Dutoit & P. Romanski. New York: Fordham University Press

Desert. (2011) Available online: https://theanarchistlibrary.org/library/anonymous-desert [Accessed 28/11/2024]

Francis, P. (2015) *Laudato Si': On Care for Our Common Home*. London: Catholic Truth Society

Hallam, R. (2019) "The Civil Resistance Model," in *This Is Not a Drill: An Extinction Rebellion Handbook*, C. Farrell, A. Green & S. Knights (Eds.). London: Penguin Books

Harrison, E. (2022) "Just Stop Oil Protesters Throw Tomato Soup on Van Gogh's Sunflowers at the National Gallery," *The Independent*. Available online: www.independent.co.uk/arts-entertainment/art/news/van-gogh-sunflowers-tomato-soup-protest-just-stop-oil-b2202765.html [Accessed 28/11/2024]

Hume, D. (1902) *An Enquiry Concerning Human Understanding*. Available online: www.gutenberg.org/files/9662/9662-h/9662-h.htm [Accessed 28/11/2024]

The Invisible Committee. (2015) *To Our Friends*. Trans. R. Hurley. Los Angeles: Semiotext(e)

Kleinberg, E. (2017) *Haunting History: For a Deconstructive Approach to the Past*. Stanford: Stanford University Press

Kropotkin, P. (2009) *Mutual Aid: A Factor of Evolution*. London: Freedom Press

Kropotkin, P. (1976) *The Essential Kropotkin*. Trans. E. Capouya & K. Tompkins. London: The Macmillan Press LTD

The Lion King. (1994) Dir. Roger Allers and Rob Minkoff. Burbank: Walt Disney Feature Animation

Malm, A. (2020) *How to Blow Up a Pipeline: Learning to Fight in a World on Fire.* London: Verso

Martel, J.R. (2014) *The One and Only Law: Walter Benjamin and the Second Commandment.* Ann Arbor: University of Michigan Press

Mary Nardini Gang (MNG). (2019) *Be Gay Do Crime.* Available online: https://the anarchistlibrary.org/library/mary-nardini-gang-be-gay-do-crime [Accessed 28/11/2024]

Nevo, I. (1995) "Difference in Context: The Logic of Antinomian Arguments," *Iyyun: The Jerusalem Philosophical Quarterly* 44, pp. 293–319

Pepperell, N. (2014) "Impure Inheritances: Spectral Materiality in Derrida and Marx," in *Messianic Thought Outside Theology,* A. Glazova & P. North (Eds.). New York: Fordham University Press

Planet of the Apes. (1968) Dir. Franklin J. Schaffner. Los Angeles: APJAC Productions

Rosenzweig, F. (1972) *The Star of Redemption.* Trans. W.H. Hallo. Boston: Beacon Press

Ruin, H. (2018) *Being with the Dead: Burial, Ancestral Politics, and the Roots of Historical Consciousness.* Standford: Stanford University Press

Schmitt, C. (2005) *Political Theology: Four Chapters on the Concept of Sovereignty.* Trans. G. Schwab. Chicago: University of Chicago Press

Secret, T. (2015) *The Politics and Pedagogy of Mourning: On Responsibility in Eulogy.* London: Bloomsbury Publishing

The Sixth Sense. (1999) Dir. M. Night Shyamalan. Burbank: Hollywood Pictures Company

Smith, M. (2000) "Environmental Antinomianism: The Moral World Turned Upside Down?" *Ethics and Environment* 5(1), pp. 125–139

Sowa, J. (2011) *Fantomowe Ciało Króla: Peryferyjne Zmagania z Nowoczesną Formą.* Kraków: TAiWPN Universitas

Tanaka, M. (2014) *Apocalypse in Contemporary Japanese Science Fiction.* London: Palgrave Macmillan

Taubes, J. (2004) *The Political Theology of Paul.* Trans. D. Hollander. Stanford: Stanford University Press

Trotsky, L. (2007) *My Life: An Attempt at an Autobiography.* New York: Dover Publications

Tsing, A.L., Bubandt, N., Gan, E., Swanson, H.A. (2017) *Arts of Living on a Damaged Planet: Ghosts and Monsters of the Anthropocene.* Minneapolis: University of Minnesota Press

INDEX